T0353493

Monographs in

Number Theory

Volume 10

Recent Progress on Topics of Ramanujan Sums and Cotangent Sums Associated with the Riemann Hypothesis

Monographs in Number Theory

ISSN 1793-8341

Monographs in

Number Theory

Volume 10

Recent Progress on Topics of Ramanujan Sums and Cotangent Sums Associated with the Riemann Hypothesis

Helmut Maier
University of Ulm, Germany

Michael Th Rassias
Hellenic Military Academy, Greece

László Tóth
University of Pécs, Hungary

World Scientific

NEW JERSEY · LONDON · SINGAPORE · BEIJING · SHANGHAI · HONG KONG · TAIPEI · CHENNAI

Published by

World Scientific Publishing Co. Pte. Ltd.

5 Toh Tuck Link, Singapore 596224

USA office: 27 Warren Street, Suite 401-402, Hackensack, NJ 07601

UK office: 57 Shelton Street, Covent Garden, London WC2H 9HE

Library of Congress Cataloging-in-Publication Data

Names: Maier, Helmut, 1953–   author. | Rassias, Michael Th., 1987–   author. |
    Tóth, László (Mathematician), author.
Title: Recent progress on topics of Ramanujan sums and cotangent sums associated with the
    Riemann hypothesis / Helmut Maier (University of Ulm, Germany), Michael Th Rassias
    (Hellenic Military Academy, Greece), László Tóth (University of Pécs, Hungary).
Description: New Jersey : World Scientific, 2022. | Series: Monographs in number theory,
    1793-8341 ; volume 10 | Includes bibliographical references and index.
Identifiers: LCCN 2021046880 | ISBN 9789811246883 (hardcover) |
    ISBN 9789811246890 (ebook for institutions) | ISBN 9789811246906 (ebook for individuals)
Subjects: LCSH: Trigonometric sums. | Mathematical analysis. | Number theory. |
    Riemann hypothesis.
Classification: LCC QA246.8.T75 M35 2022 | DDC 512.7/3--dc23/eng/20211102
LC record available at https://lccn.loc.gov/2021046880

British Library Cataloguing-in-Publication Data
A catalogue record for this book is available from the British Library.

For any available supplementary material, please visit
https://www.worldscienti ic.com/worldscibooks/10.1142/12544#t=suppl

Desk Editors: Vishnu Mohan/Rok Ting Tan

Typeset by Stallion Press
Email: enquiries@stallionpress.com

Printed in Singapore

Preface

In this monograph, we study recent results on some categories of trigonometric/exponential sums along with various of their applications in Mathematical Analysis and Analytic Number Theory. Through the two chapters of this monograph, we wish to highlight the applicability and breadth of techniques of trigonometric/exponential sums in various problems focusing on the interplay of Mathematical Analysis and Analytic Number Theory. We wish to stress the point that the goal is not only to prove the desired results, but also to present a plethora of intermediate propositions and corollaries investigating the behavior of such sums, which can also be applied in completely different problems and settings than the ones treated within this monograph. In the present work, we mainly focus on the applications of trigonometric/exponential sums in the study of Ramanujan sums — which constitute a very classical domain of research in Number Theory — as well as the study of certain cotangent sums with a wide range of applications, especially in the study of Dedekind sums and a facet of the research conducted on the Riemann Hypothesis. For example, in our study of the cotangent sums treated within the second chapter, the methods and techniques employed reveal unexpected connections with independent and very interesting problems investigated in the past by R. de la Bretèche and G. Tenenbaum on trigonometric series, as well as by S. Marmi, P. Moussa and J.-C. Yoccoz on Dynamical Systems.

More specifically, in the first chapter, we present an overview of classical and very recent results on various properties and applications of Ramanujan sums and their analogs as well as generalizations. In particular, we discuss properties of even functions (mod r), the discrete Fourier transform (DFT) of even functions (mod r), Menon-type identities and certain results for the number of solutions of linear congruences with constraints, respectively quadratic congruences in several variables. Also, we present results on expansions of arithmetic functions of several variables with respect to Ramanujan sums and their analogs.

In the second chapter, we give an overview of recent results on various aspects of a specific category of cotangent sums which appear in the investigation of the Nyman–Beurling criterion for the Riemann Hypothesis. As the background needed to follow parts of this chapter may be more demanding for some readers, we omit certain technical details (providing though the corresponding references for further investigation) and in those parts we focus on the elaboration of the central ideas and steps, featuring all the crucial lemmas, propositions and corollaries, certain of which the reader may find interesting as independent results in the domain of trigonometric/ exponential sums.

Overall, a reader who has mastered fundamentals of Mathematical Analysis as well as has a working knowledge of Classical and Analytic Number Theory will be able to gradually follow all the parts of the monograph. Therefore, it is hoped that the present monograph will be of interest to advanced undergraduate and graduate students as well as researchers who wish to be informed on the latest developments on the topics treated.

We would like to express our thanks to the staff of World Scientific for our excellent collaboration for the production of this book.

Helmut Maier
Ulm, Germany

Michael Th. Rassias
Athens, Greece

László Tóth
Pécs, Hungary

Contents

Chapter 1

Ramanujan Sums: A Survey of Recent Results

1.1. Preliminaries

In this chapter, we present an overview of the results by Haukkanen and Tóth [39], Tóth [108, 111–115, 117–120], Tóth and Haukkanen [121] on various properties and applications of Ramanujan sums and some of their analogs and generalizations. The Ramanujan sum $c_q(n)$ is defined as the sum of nth powers of the primitive qth roots of unity, where $n \in \mathbb{Z}$, $q \in \mathbb{N} = \{1, 2, \ldots\}$. That is,

$$c_q(n) = \sum_{\substack{k=1 \\ (k,q)=1}}^{q} e(kn/q),$$

where $e(x) = e^{2\pi i x}$. These sums were studied by Srinivasa Ramanujan [92] in 1918. His principal object was to deduce series expansions

$$f(n) = \sum_{q=1}^{\infty} a_q c_q(n) \tag{1.1.1}$$

for special arithmetic functions f. For example, Ramanujan [92] proved for the sum of divisors function $\sigma(n)$ that

$$\frac{\sigma(n)}{n} = \frac{\pi^2}{6} \sum_{q=1}^{\infty} \frac{c_q(n)}{q^2}$$

$$= \frac{\pi^2}{6} \left(1 + \frac{(-1)^n}{2^2} + \frac{2\cos(2\pi n/3)}{3^2} + \frac{2\cos(\pi n/2)}{4^2} + \cdots \right), \tag{1.1.2}$$

1

for every $n \in \mathbb{N}$, which shows how the values of $\sigma(n)/n$ fluctuate harmonically about their mean value $\pi^2/6$.

Other related results of Ramanujan [92] are

$$d(n) = -\sum_{q=1}^{\infty} \frac{\log q}{q} c_q(n), \qquad (1.1.3)$$

where $d(n)$ denotes the number of divisors of n,

$$\sum_{q=1}^{\infty} \frac{c_q(n)}{q} = 0 \quad (n \in \mathbb{N}),$$

which is equivalent to the prime number theorem, and

$$\sum_{n=1}^{\infty} \frac{c_q(n)}{n} = -\Lambda(q) \quad (q > 1), \qquad (1.1.4)$$

where Λ is the von Mangoldt function, and the sum is over the other variable (namely n).

There are many results in the literature on various aspects of Ramanujan expansions of arithmetic functions of one variable. See, e.g., the book by Schwarz and Spilker [100] and the survey papers by Lucht [59] and Ram Murty [93].

Tóth [118] investigated Ramanujan expansions of arithmetic functions of several variables. For example, as special cases of more general results, it was proved in [118] that for every $n_1, \ldots, n_k \in \mathbb{N}$ the following series are absolutely convergent:

$$\frac{\sigma((n_1, \ldots, n_k))}{(n_1, \ldots, n_k)} = \zeta(k+1) \sum_{q_1, \ldots, q_k=1}^{\infty} \frac{c_{q_1}(n_1) \cdots c_{q_k}(n_k)}{[q_1, \ldots, q_k]^{k+1}} \quad (k \geq 1), \quad (1.1.5)$$

$$d((n_1, \ldots, n_k)) = \zeta(k) \sum_{q_1, \ldots, q_k=1}^{\infty} \frac{c_{q_1}(n_1) \cdots c_{q_k}(n_k)}{[q_1, \ldots, q_k]^k} \quad (k \geq 2), \qquad (1.1.6)$$

where ζ is the Riemann zeta function, $(n_1, \ldots, n_k)$ denotes the greatest common divisor of $n_1, \ldots, n_k$ and $[q_1, \ldots, q_k]$ stands for the least common multiple of $q_1, \ldots, q_k$.

For $k = 1$, identity (1.1.5) recovers (1.1.2). However, (1.1.6) does not have a direct one-dimensional analog. Identity (1.1.3) cannot be obtained by the same approach.

Expansions of arithmetic functions of several variables with respect to the unitary Ramanujan sums $c_q^*(n)$ and certain modified unitary Ramanujan sums, respectively were also considered by Tóth [118, 119].

Ramanujan sums occur in various other formulas. For example, let

$$m_1, \ldots, m_r, M \in \mathbb{N}, \quad m = [m_1, \ldots, m_r], \quad m \mid M.$$

Then the number of incongruent solutions $(x_1, \ldots, x_r) \in \mathbb{Z}^r$ of the linear congruence

$$x_1 + \cdots + x_r \equiv 0 \pmod{M}$$

satisfying

$$(x_1, M) = M/m_1, \ldots, (x_r, M) = M/m_r$$

is given by

$$E(m_1, \ldots, m_r) = \frac{1}{M} \sum_{k=1}^{M} c_{m_1}(k) \cdots c_{m_r}(k). \tag{1.1.7}$$

Here $E(m_1, \ldots, m_r)$ is called the "orbicyclic" arithmetic function. It also has combinatorial and topological applications. See Mednykh and Nedela [76], Liskovets [58]. Tóth [113] pointed out that for any $m_1, \ldots, m_r \in \mathbb{N}$,

$$E(m_1, \ldots, m_r) = \sum_{d_1 \mid m_1, \ldots, d_r \mid m_r} \frac{d_1 \cdots d_r}{[d_1, \ldots, d_r]} \mu(m_1/d_1) \cdots \mu(m_r/d_r),$$

$$\tag{1.1.8}$$

where μ is the Möbius function. This identity shows that $E(m_1, \ldots, m_r)$ is a multiplicative function of r variables, being the convolution of multiplicative functions.

As another example, let $N_k(n, r)$ denote the number of incongruent solutions $(x_1, \ldots, x_k) \in \mathbb{Z}^k$ of the quadratic congruence

$$x_1^2 + \cdots + x_k^2 \equiv n \pmod{r}.$$

The function $r \mapsto N_k(n, r)$ is multiplicative. Therefore, it is sufficient to consider the case $r = p^s$, a prime power. Identities for $N_k(n, p^s)$ can be derived using Gauss and Jacobi sums. It is less known that for k even and r odd, $N_k(n, r)$ can be expressed in a compact form using Ramanujan's sum.

Namely, one has the following identities: If $k, r \in \mathbb{N}$, $k \equiv 0 \pmod 4$, r is odd and $n \in \mathbb{Z}$, then

$$N_k(n,r) = r^{k-1} \sum_{d|r} \frac{c_d(n)}{d^{k/2}}, \qquad (1.1.9)$$

and if $k \equiv 2 \pmod 4$, r odd and $n \in \mathbb{Z}$, then

$$N_k(n,r) = r^{k-1} \sum_{d|r} (-1)^{(d-1)/2} \frac{c_d(n)}{d^{k/2}}, \qquad (1.1.10)$$

which are special cases of certain more general identities deduced by Cohen [27], of which proofs are lengthy and use Cohen's previous work.

Tóth [114] gave short direct proofs of identities concerning $N_k(n,r)$ for every $k, r \in \mathbb{N}$, $n \in \mathbb{Z}$, including (1.1.9) and (1.1.10). Also, paper [114] contains some more historical remarks on this topic, and presents asymptotic formulas for the sums

$$\sum_{r \leq x} N_k(n,r),$$

for some special values of k and n, not given in the literature.

If $n \in \mathbb{N}$ and $s \in \mathbb{Z}$, then

$$\sum_{\substack{k=1 \\ (k,n)=1}}^{n} c_n(k-s) = \mu(n) c_n(s), \qquad (1.1.11)$$

where μ is the Möbius function (see [75, Chapter 2]). The shape of this identity is similar to that of Menon's formula [77], stating that

$$\sum_{\substack{a=1 \\ (a,n)=1}}^{n} (a-1,n) = \phi(n) d(n) \quad (n \in \mathbb{N}), \qquad (1.1.12)$$

where ϕ is Euler's arithmetic function.

A common generalization of identities (1.1.11) and (1.1.12), involving even functions (mod n) and Dirichlet characters (mod n) was obtained by Tóth [117].

It is not surprising that Ramanujan sums are related to the cyclotomic polynomials $\Phi_n(x)$. For every $n > 1$ and $x \in \mathbb{C}$, $|x| < 1$ one has

$$\Phi_n(x) = \exp\left(-\sum_{k=1}^{\infty} \frac{c_n(k)}{k} x^k\right), \tag{1.1.13}$$

and for $n \geq 1$,

$$\frac{\Phi_n'\left(\frac{1}{x}\right)}{\Phi_n\left(\frac{1}{x}\right)} = \frac{x}{1-x^n} \sum_{k=0}^{n-1} c_n(k) x^k, \tag{1.1.14}$$

as shown by Nicol [86] and Motose [81].

Tóth [111] investigated the polynomials with integer coefficients

$$R_n(x) = \sum_{k=0}^{n-1} c_n(k) x^k \tag{1.1.15}$$

appearing in (1.1.14) and deduced for $R_n(x)$ identities that are similar to the well known identities valid for the cyclotomic polynomials $\Phi_n(x)$.

The sums $c_q(n)$ also play an important role in the proof of Vinogradov's theorem concerning the number of representations of an odd integer as the sum of three primes (see, e.g., [85, Chapter 8]).

This chapter is organized as follows. Some basic properties of arithmetic functions of one and several variables are presented in Section 1.2. Basic facts on Ramanujan sums are included in Section 1.3. Section 1.4 concerns sums of products of Ramanujan sums and generalizations of the orbicyclic function (1.1.7). Certain weighted averages of Ramanujan sums are considered in Section 1.5. Properties of even functions (mod r), the discrete Fourier transform (DFT) of even functions (mod r) and Menon-type identities are discussed in Section 1.6. Results for the number of solutions of linear congruences with constraints, respectively quadratic congruences in several variables are presented in Section 1.7. Results on the polynomials (1.1.15) are given in Section 1.8. Some analogs and generalizations of Ramanujan sums are objects of Section 1.9. Finally, results on expansions of arithmetic functions of several variables with respect to Ramanujan sums and their analogs are given in Section 1.10.

For general accounts on the discussed topics we refer to the books by Apostol [4], McCarthy [75], Montgomery [78], Montgomery and Vaughan [79], Narkiewicz [84], Schwarz and Spilker [100].

1.2. Arithmetic Functions of One and Several Variables

Properties of (multiplicative) arithmetic functions of one variable, i.e., functions $f \colon \mathbb{N} \to \mathbb{C}$ (satisfying $f(mn) = f(m)f(n)$ whenever $(m,n) = 1$) are well known in the literature. Less known are (multiplicative) arithmetic functions of several variables. We outline the basic facts we need on arithmetic functions of r variables. See the survey by Tóth [116] for a more complete discussion. If $r = 1$, then we recover properties of the functions of a single variable.

Let $r \in \mathbb{N}$ be fixed. The set $\mathcal{A}_r$ of arithmetic functions $f \colon \mathbb{N}^r \to \mathbb{C}$ of r variables is a $\mathbb{C}$-linear space with the usual linear operations. With the Dirichlet convolution defined by

$$(f * g)(n_1, \ldots, n_r) = \sum_{d_1 \mid n_1, \ldots, d_r \mid n_r} f(d_1, \ldots, d_r) g(n_1/d_1, \ldots, n_r/d_r)$$

the space $\mathcal{A}_r$ forms a unital commutative $\mathbb{C}$-algebra. The unity is the function ε_r defined by

$$\varepsilon_r(1, \ldots, 1) = 1 \quad \text{and} \quad \varepsilon_r(n_1, \ldots, n_r) = 0 \quad \text{for} \quad n_1 \cdots n_r > 1.$$

The group of invertible functions is

$$\mathcal{A}_r^{(1)} = \{ f \in \mathcal{A}_r : f(1, \ldots, 1) \neq 0 \}.$$

The inverse of the constant 1 function is the r variables Möbius function μ_r, given by

$$\mu_r(n_1, \ldots, n_r) = \mu(n_1) \cdots \mu(n_r),$$

where μ is the classical Möbius function.

A function $f \in \mathcal{A}_r$ is called multiplicative if f is not identically zero and

$$f(m_1 n_1, \ldots, m_r n_r) = f(m_1, \ldots, m_r) f(n_1, \ldots, n_r)$$

holds for any $m_1, \ldots, m_r, n_1, \ldots, n_r \in \mathbb{N}$ such that $(m_1 \cdots m_r, n_1 \cdots n_r) = 1$.

We use for the prime power factorization of $n \in \mathbb{N}$ the notation

$$n = \prod_p p^{\nu_p(n)},$$

the product being over the primes p, where all but a finite number of the exponents $\nu_p(n)$ are zero.

If f is multiplicative, then it is determined by the values

$$f(p^{\nu_1}, \ldots, p^{\nu_r}),$$

where p is prime and $\nu_1, \ldots, \nu_r \geq 0$. More specifically,

$$f(1, \ldots, 1) = 1$$

and for any $n_1, \ldots, n_r \in \mathbb{N}$,

$$f(n_1, \ldots, n_r) = \prod_p f(p^{\nu_p(n_1)}, \ldots, p^{\nu_p(n_r)}).$$

For example, the functions

$$(n_1, \ldots, n_r) \mapsto \gcd(n_1, \ldots, n_r)$$

and

$$(n_1, \ldots, n_r) \mapsto \mathrm{lcm}(n_1, \ldots, n_r)$$

are multiplicative. The Dirichlet convolution preserves the multiplicativity of functions. More exactly, the set $\mathcal{M}_r$ of multiplicative functions of r variables is a subgroup of $\mathcal{A}_r^{(1)}$ with respect to the Dirichlet convolution.

A divisor d of n is called a unitary divisor (or block divisor) if $(d, n/d) = 1$, notation $d \parallel n$. Note that this is in agreement with the standard notation $p^a \parallel n$ used for prime powers p^a. If the prime power factorization of n is

$$n = p_1^{a_1} \cdots p_s^{a_s},$$

then its unitary divisors are

$$d = p_1^{b_1} \cdots p_s^{b_s},$$

where $b_i = 0$ or $b_i = a_i$ for any $1 \leq i \leq s$. If n is squarefree, then all its divisors are unitary divisors.

The linear space $\mathcal{A}_r$ forms another unital commutative $\mathbb{C}$-algebra with the unitary convolution defined by

$$(f \times g)(n_1, \ldots, n_r) = \sum_{d_1 \parallel n_1, \ldots, d_r \parallel n_r} f(d_1, \ldots, d_r) g(n_1/d_1, \ldots, n_r/d_r).$$

$$(1.2.1)$$

Here the unity is the function ε_r again. Note that $(\mathcal{A}_r, +, \times)$ is not an integral domain, since there exist divisors of zero. The group of invertible functions is again $\mathcal{A}_r^{(1)}$. The unitary r variables Möbius function μ_r^* is defined as the inverse, with respect to (1.2.1), of the constant 1 function. One has

$$\mu_r^*(n_1, \ldots, n_r) = (-1)^{\omega(n_1) + \cdots + \omega(n_r)},$$

where $\omega(n)$ denotes the number of distinct prime factors of n.

The set $\mathcal{M}_r$ of multiplicative functions of r variables is a subgroup of $\mathcal{A}_r^{(1)}$ with respect to the unitary convolution.

1.3. Basic Properties of Ramanujan Sums

A useful representation of the Ramanujan sums is

$$c_q(n) = \sum_{d \mid (q,n)} d\mu(q/d) \quad (n \in \mathbb{Z}, q \in \mathbb{N}), \tag{1.3.1}$$

that is, $c_.(n) = \mu * \eta_.(n)$, where

$$\eta_q(n) = \begin{cases} q & \text{if } q \mid n, \\ 0 & \text{otherwise.} \end{cases} \tag{1.3.2}$$

A direct consequence of (1.3.1) is that all values of $c_q(n)$ are integers. If $q \mid n$, then $c_q(n) = \phi(q)$, Euler's totient function. If $(q,n) = 1$, then $c_q(n) = \mu(q)$, the Möbius function. For any fixed $n \in \mathbb{N}$, the function $q \mapsto c_q(n)$ is multiplicative and for any prime power p^a $(a \in \mathbb{N})$,

$$c_{p^a}(n) = \begin{cases} p^a - p^{a-1} & \text{if } p^a \mid n, \\ -p^{a-1} & \text{if } p^{a-1} \mid n, p^a \nmid n, \\ 0 & \text{if } p^{a-1} \nmid n. \end{cases} \tag{1.3.3}$$

On the other hand, $n \mapsto c_q(n)$ is multiplicative if and only if $\mu(q) = 1$. This follows from the identity

$$c_q(m) c_q(n) = \mu(q) c_q(mn), \tag{1.3.4}$$

valid for $(m, n) = 1$ (see [75, p. 90]). Also, for fixed q, the function $n \mapsto c_q(n)$ is an even function (mod q), its values depend only on the gcd(q, n).

The Hölder evaluation of the Ramanujan sums is given by

$$c_q(n) = \frac{\phi(q)\mu(q/(q,n))}{\phi(q/(q,n))} \quad (n \in \mathbb{Z}, q \in \mathbb{N}). \tag{1.3.5}$$

We have

$$\frac{1}{q}\sum_{n=1}^{q} c_q(n) = \begin{cases} 1 & \text{if } q = 1, \\ 0 & \text{otherwise,} \end{cases}$$

and if $[q_1, q_2] \mid N$, then

$$E(q_1, q_2) = \frac{1}{N}\sum_{n=1}^{N} c_{q_1}(n)c_{q_2}(n)$$

$$= \begin{cases} \phi(q) & \text{if } q_1 = q_2 = q, \\ 0 & \text{otherwise.} \end{cases} \tag{1.3.6}$$

A short direct proof of the orthogonality property (1.3.6) is as follows. [113, Appl. 8]. By using (1.3.1) we deduce that

$$E(q_1, q_2) = \frac{1}{N}\sum_{n=1}^{N} \sum_{d_1|(n,q_1)} d_1\mu(q_1/d_1) \sum_{d_2|(n,q_2)} d_2\mu(q_2/d_2)$$

$$= \frac{1}{N}\sum_{d_1|q_1, d_2|q_2} d_1\mu(q_1/d_1)d_2\mu(q_2/d_2) \sum_{\substack{n=1 \\ d_1|n, d_2|n}}^{N} 1,$$

where the inner sum is

$$\sum_{\substack{n=1 \\ [d_1,d_2]|n}}^{N} 1 = N/[d_1, d_2].$$

We find that

$$E(q_1, q_2) = \sum_{d_1|q_1, d_2|q_2} \frac{d_1 d_2 \mu(q_1/d_1)\mu(q_2/d_2)}{[d_1, d_2]}$$

$$= \sum_{d_1|q_1, d_2|q_2} (d_1, d_2)\mu(q_1/d_1)\mu(q_2/d_2). \tag{1.3.7}$$

Note that $E(q_1, q_2)$ is the two variables special case of the orbicyclic function, defined by (1.1.7). The general identity (1.1.8) is obtained in a way similar to (1.3.7).

Now (1.3.7) and the Gauss formula

$$\sum_{d|n} \phi(d) = n$$

give

$$E(q_1, q_2) = \sum_{d_1|q_1, d_2|q_2} \mu(q_1/d_1)\mu(q_2/d_2) \sum_{\delta|(d_1, d_2)} \phi(\delta)$$

$$= \sum_{\delta a k = q_1, \delta b \ell = q_2} \mu(k)\mu(\ell)\phi(\delta)$$

$$= \sum_{\delta u = q_1, \delta v = q_2} \phi(\delta) \sum_{ak=u} \mu(k) \sum_{b\ell=v} \mu(\ell),$$

where one of the inner sums is zero, unless $u = v = 1$.

We obtain that

$$E(q_1, q_2) = \begin{cases} \phi(q) & \text{for } q_1 = q_2 = q, \\ 0 & \text{otherwise.} \end{cases}$$

This proves (1.3.6).

Let

$$M(f) = \lim_{x \to \infty} \frac{1}{x} \sum_{n \le x} f(n)$$

denote the mean value of the arithmetic function f, if the limit exists.

We also have the orthogonality relations

$$M\left(c_{q_1}(.)c_{q_2}(.)\right) = \begin{cases} \phi(q) & \text{if } q_1 = q_2 = q, \\ 0 & \text{otherwise,} \end{cases} \qquad (1.3.8)$$

which suggests that certain functions f have expansions of type (1.1.1).

Identity (1.3.1) also shows that $c_q(n)$ is multiplicative, viewed as a function of two variables, being the convolution of the multiplicative functions f and g defined by

$$f(n_1, n_2) = \begin{cases} n & \text{for } n_1 = n_2 = n, \\ 0 & \text{for } n_1 \neq n_2, \end{cases}$$

and

$$g(n_1, n_2) = \mu(n_1) \quad \text{for every } n_1, n_2 \in \mathbb{N}.$$

The following properties can be used to prove results on the expansions of arithmetic functions with respect to Ramanujan sums. For any $q, n \in \mathbb{N}$,

$$\sum_{d|q} |c_d(n)| = 2^{\omega(q/(n,q))}(n, q), \tag{1.3.9}$$

$$\sum_{d|q} |c_d(n)| \leq 2^{\omega(q)} n. \tag{1.3.10}$$

1.4. Sums of Products of Ramanujan Sums

Formulas (1.1.7) and (1.1.11) suggest the consideration of the following generalization. Let $m_1, \ldots, m_r \in \mathbb{N}$ ($r \in \mathbb{N}$) and $m = \operatorname{lcm}(m_1, \ldots, m_r)$. Let $G = (g_1, \ldots, g_r)$ be a system of polynomials with integer coefficients. Define

$$E_G(m_1, \ldots, m_r) = \frac{1}{m} \sum_{k=1}^{m} c_{m_1}(g_1(k)) \cdots c_{m_r}(g_r(k)), \tag{1.4.1}$$

$$R_G(m_1, \ldots, m_r) = \sum_{\substack{k=1 \\ \gcd(k,m)=1}}^{m} c_{m_1}(g_1(k)) \cdots c_{m_r}(g_r(k)), \tag{1.4.2}$$

where we can assume that $m_i > 1$ ($1 \leq i \leq r$), since $c_1(k) = 1$ for any $k \in \mathbb{Z}$.

If

$$g_1(x) = \cdots = g_r(x) = x,$$

then (1.4.1) reduces to the function (1.1.7). In the one variable case, i.e., $r = 1$, and selecting the linear polynomial $g_1(x) = x - s$, (1.4.2) gives the sum in (1.1.11).

We have the following general results. Let $N_G(m_1, \ldots, m_r)$ denote the number of solutions x (mod $\operatorname{lcm}(m_1, \ldots, m_r)$) of the simultaneous congruences

$$g_1(x) \equiv 0 \ (\text{mod } m_1), \ldots, g_r(x) \equiv 0 \ (\text{mod } m_r). \tag{1.4.3}$$

Furthermore, let $\eta_G(m_1, \ldots, m_r)$ denote the number of solutions

$$x \ (\text{mod } \text{lcm}(m_1, \ldots, m_r))$$

of (1.4.3) such that

$$\gcd(x, m_1) = 1, \ldots, \gcd(x, m_r) = 1.$$

Here both $N_G(m_1, \ldots, m_r)$ and $\eta_G(m_1, \ldots, m_r)$ are multiplicative functions of r variables.

Theorem 1.4.1 ([112, Theorem 1]). *If G is an arbitrary system of polynomials with integer coefficients, then for any $m_1, \ldots, m_r \in \mathbb{N}$,*

$$E_G(m_1, \ldots, m_r) = \sum_{d_1 | m_1, \ldots, d_r | m_r} \frac{d_1 \mu(m_1/d_1) \cdots d_r \mu(m_r/d_r)}{\text{lcm}(d_1, \ldots, d_r)}$$

$$\times N_G(d_1, \ldots, d_r), \tag{1.4.4}$$

representing an integer-valued multiplicative function.

Theorem 1.4.2 ([112, Theorem 2]). *If G is an arbitrary system of polynomials with integer coefficients, then for any $m_1, \ldots, m_r \in \mathbb{N}$,*

$$R_G(m_1, \ldots, m_r) = \phi(m) \sum_{d_1 | m_1, \ldots, d_r | m_r} \frac{d_1 \mu(m_1/d_1) \cdots d_r \mu(m_r/d_r)}{\phi(\text{lcm}(d_1, \ldots, d_r))}$$

$$\times \eta_G(d_1, \ldots, d_r), \tag{1.4.5}$$

representing a multiplicative function.

We present here the following consequences.

Corollary 1.4.3 ([112, Corollary 3]). *For every $m_1, m_2 \in \mathbb{N}$ with $m = \text{lcm}(m_1, m_2)$ and every $a_1, a_2 \in \mathbb{Z}$,*

$$E_{(a_1, a_2)}(m_1, m_2) := \frac{1}{m} \sum_{k=1}^{m} c_{m_1}(k - a_1) c_{m_r}(k - a_2)$$

$$= \sum_{\substack{d_1 | m_1, d_2 | m_2 \\ \gcd(d_1, d_2) | a_1 - a_2}} \gcd(d_1, d_2) \mu(m_1/d_1) \mu(m_2/d_2). \tag{1.4.6}$$

Furthermore, if $|a_1 - a_2| = 1$, then

$$E_{(a_1,a_2)}(m_1, m_2) = \begin{cases} (-1)^{\omega(m)} & \text{if } m_1 = m_2 = m \text{ is squarefree,} \\ 0 & \text{otherwise,} \end{cases} \quad (1.4.7)$$

$\omega(m)$ *denoting the number of distinct prime factors of m.*

Corollary 1.4.4 ([112, **Corollary 10**]). *If $m_1, m_2 \in \mathbb{N}$ with $m = \mathrm{lcm}(m_1, m_2)$ and $a_1, a_2 \in \mathbb{Z}$, then*

$$R_{(a_1,a_2)}(m_1, m_2) := \sum_{\substack{k=1 \\ \gcd(k,m)=1}}^{m} c_{m_1}(k - a_1) c_{m_2}(k - a_2)$$

$$= \phi(m) \sum_{\substack{d_1|m_1, d_2|m_2 \\ \gcd(d_1,a_1)=1, \gcd(d_2,a_2)=1 \\ \gcd(d_1,d_2)|a_1-a_2}} \frac{d_1 \mu(m_1/d_1) d_2 \mu(m_2/d_2)}{\phi(\mathrm{lcm}(d_1, d_2))}.$$

$$(1.4.8)$$

Furthermore, if $\gcd(a_1, m_1) = \gcd(a_2, m_2) = 1$ and $|a_1 - a_2| = 1$, then

$$R_{(a_1,a_2)}(m_1, m_2) = \begin{cases} (-1)^{\omega(\gcd(m_1,m_2))} & \text{if } m_1 \text{ and } m_2, \\ \psi(\gcd(m_1, m_2)) & \text{are squarefree} \quad (1.4.9) \\ 0 & \text{otherwise,} \end{cases}$$

where

$$\psi(n) = n \prod_{p|n} \left(1 + \frac{1}{p}\right)$$

is the Dedekind function.

Several other special cases can be discussed. For further results, see paper [112].

1.5. Weighted Averages of Ramanujan Sums

Alkan [2] considered the weighted average of the Ramanujan sums $c_k(j)$ defined by

$$S_r(k) = \frac{1}{k^{r+1}} \sum_{j=1}^{k} j^r c_k(j) \quad (r \in \mathbb{N}), \quad (1.5.1)$$

being motivated by the use of (1.5.1) in proving exact formulas for certain mean square averages of special values of L-functions. He showed that for every $k, r \in \mathbb{N}$,

$$S_r(k) = \frac{\phi(k)}{2k} + \frac{1}{r+1} \sum_{m=1}^{\lfloor r/2 \rfloor} \binom{r+1}{2m} B_{2m} \prod_{p|k} \left(1 - \frac{1}{p^{2m}}\right), \qquad (1.5.2)$$

leading to the asymptotic formula

$$\sum_{k \leq x} S_r(k) = \left(\frac{3}{\pi^2} + \frac{1}{r+1} \sum_{m=1}^{\lfloor r/2 \rfloor} \binom{r+1}{2m} \frac{B_{2m}}{\zeta(2m+1)}\right) x + O(\log x),$$

where B_m ($m \geq 0$) are the Bernoulli numbers and ζ is the Riemann zeta function (see [2, Eq. (2.19), Theorem 1]. The same identity (1.5.2) and the same proof, based on Hölder's evaluation of the Ramanujan sums were given by Alkan also in [3, Proof of Theorem 1].

Tóth [115] presented a simpler proof of identity (1.5.2), and established identities for other weighted averages of the Ramanujan sums with weights concerning logarithms, values of arithmetic functions for gcd's, the Gamma function, the Bernoulli polynomials and binomial coefficients.

Proposition 1.5.1 (Tóth [115, Propositions 2, 4 and 5]). *For every* $k \in \mathbb{N}$,

$$\frac{1}{k} \sum_{j=1}^{k} (\log j) c_k(j) = \Lambda(k) + \sum_{d|k} \frac{\mu(d)}{d} \log(d!), \qquad (1.5.3)$$

where Λ *is the von Mangoldt function,*

$$\frac{1}{\phi(k)} \sum_{j=1}^{k} (\log \Gamma(j/k)) c_k(j) = \frac{1}{2} \sum_{p|k} \frac{\log p}{p-1} - \frac{\log 2\pi}{2} \quad (k \geq 2), \qquad (1.5.4)$$

$$\frac{1}{2^k} \sum_{j=0}^{k} \binom{k}{j} c_k(j) = \sum_{d|k} \mu(k/d) \sum_{\ell=1}^{d} (-1)^{\ell k/d} \cos^k(\ell \pi/d). \qquad (1.5.5)$$

Note the symmetry property

$$\binom{k}{j}c_k(j) = \binom{k}{k-j}c_k(k-j) \quad (0 \le j \le k).$$

Proof. We prove identity (1.5.4). It is well known that for every $n \in \mathbb{N}$,

$$\prod_{k=1}^{n} \Gamma(k/n) = \frac{(2\pi)^{(n-1)/2}}{\sqrt{n}}, \tag{1.5.6}$$

which is a consequence of Gauss' multiplication formula (cf., e.g., [31, Proposition 9.6.33]). We obtain from (1.3.1) and (1.5.6) that

$$\sum_{j=1}^{k} (\log \Gamma(j/k)) c_k(j) = \sum_{j=1}^{k} (\log \Gamma(j/k)) \sum_{d \,|\, \gcd(k,j)} d\mu(k/d)$$

$$= \sum_{d \,|\, k} d\mu(k/d) \sum_{m=1}^{k/d} \log \Gamma(md/k)$$

$$= \sum_{d \,|\, k} (k/d)\mu(d) \sum_{m=1}^{d} \log \Gamma(m/d)$$

$$= \sum_{d \,|\, k} (k/d)\mu(d) \log \frac{(2\pi)^{(d-1)/2}}{\sqrt{d}}$$

$$= \log(2\pi) \sum_{d \,|\, k} \frac{k}{d}\mu(d) \frac{d-1}{2} - \frac{1}{2} \sum_{d \,|\, k} \frac{k}{d}\mu(d) \log d$$

$$= \log(2\pi) \left(\frac{k}{2} \sum_{d \,|\, k} \mu(d) - \frac{1}{2} \sum_{d \,|\, k} \frac{k}{d}\mu(d) \right)$$

$$- \frac{k}{2} \sum_{d \,|\, k} \frac{\mu(d)}{d} \log d,$$

where the first sum is zero for $k > 1$ and the second sum is $\phi(k)$.
Now the use of the identity

$$\sum_{d \,|\, k} \frac{\mu(d)}{d} \log d = -\frac{\phi(k)}{k} \sum_{p \,|\, k} \frac{\log p}{p-1},$$

(see, e.g., [31, Example 10.8.45]) completes the proof. $\qquad \square$

Generalizations and analogues of identities (1.5.3), (1.5.4) and (1.5.5) were investigated by Ikeda, Kiuchi and Matsuoka [44], Kiuchi [53–55], and Namboothiri [82]. For example, Namboothiri [82] considered Cohen's generalization of the Ramanujan sums, defined by

$$c_k^{(s)}(j) = \sum_{\substack{m=1 \\ (m,k^s)_s=1}}^{n} e(jm/n^s), \qquad (1.5.7)$$

where $(m, k^s)_s$ is the greatest common s-power divisor of m and k^s ($s \in \mathbb{N}$). Note that moments of averages of the function (1.5.7) were studied by Robles and Roy [97].

1.6. Properties of Even Functions (mod r)

1.6.1. Periodic functions (mod r)

Let $\mathcal{A} = \mathcal{A}_1$ denote the set of arithmetic functions of one variable. A function $f \in \mathcal{A}$ is called periodic (mod r) or r-periodic if $f(n + r) = f(n)$ holds for every $n \in \mathbb{N}$, where $r \in \mathbb{N}$ is a fixed number. This periodicity extends f to a function defined on $\mathbb{Z}$. The set $\mathcal{P}_r$ of r-periodic functions forms an r-dimensional subspace of $\mathcal{A}$. A function $f \in \mathcal{A}$ is said to be periodic if

$$f \in \bigcup_{r \in \mathbb{N}} \mathcal{P}_r.$$

The functions δ_k with $1 \le k \le r$ given by

$$\delta_k(n) = 1 \text{ for } n \equiv k \pmod{r}$$

and

$$\delta_k(n) = 0 \text{ for } n \not\equiv k \pmod{r}$$

form a basis of $\mathcal{P}_r$ (standard basis).

The functions e_k with $1 \le k \le r$ defined by $e_k(n) = e(kn/r)$ (additive characters) form another basis of the space $\mathcal{P}_r$. Therefore, every r-periodic function f has a Fourier expansion of the form

$$f(n) = \sum_{k(\mathrm{mod}\ r)} g(k)e(kn/r) \qquad (n \in \mathbb{N}), \tag{1.6.1}$$

where the Fourier coefficients $g(k)$ are uniquely determined and given by

$$g(n) = \frac{1}{r} \sum_{k(\mathrm{mod}\ r)} f(k)e(-kn/r) \qquad (n \in \mathbb{N}), \tag{1.6.2}$$

and the function g is also r-periodic.

For a function $f \in \mathcal{P}_r$ its discrete (finite) Fourier transform (DFT) is the function $\widehat{f} \in \mathcal{A}$ defined by

$$\widehat{f}(n) = \sum_{k(\mathrm{mod}\ r)} f(k)e(-kn/r) \qquad (n \in \mathbb{N}), \tag{1.6.3}$$

where by (1.6.3) and (1.6.2) one has $\widehat{f} = rg$.

For any $r \in \mathbb{N}$, the DFT is an automorphism of $\mathcal{P}_r$ satisfying

$$\widehat{\widehat{f}}(n) = rf(-n) \quad (n \in \mathbb{Z}).$$

The inverse discrete Fourier transform (IDFT) is given by

$$f(n) = \frac{1}{r} \sum_{k(\mathrm{mod}\ r)} \widehat{f}(k)e(kn/r) \qquad (n \in \mathbb{N}). \tag{1.6.4}$$

If $f \in \mathcal{P}_r$, then

$$\sum_{n=1}^{r} |\widehat{f}(n)|^2 = r \sum_{n=1}^{r} |f(n)|^2, \tag{1.6.5}$$

which is a version of Parseval's formula.

Let $f, h \in \mathcal{P}_r$. The Cauchy convolution of f and h is defined by

$$(f \otimes h)(n) = \sum_{a(\mathrm{mod}\ r)} f(a)h(n-a) \qquad (n \in \mathbb{N}), \tag{1.6.6}$$

where $(\mathcal{P}_r, \otimes)$ is a unital commutative semigroup, the unity being the function $\varepsilon^{(r)}$ given by $\varepsilon^{(r)}(n) = 1$ for $r \mid n$ and $\varepsilon^{(r)}(n) = 0$ otherwise. Also,

$$\widehat{f \otimes h} = \widehat{f}\,\widehat{h} \quad \text{and} \quad \widehat{f} \otimes \widehat{h} = r\widehat{fh}.$$

1.6.2. Even functions (mod r)

A function $f \in \mathcal{A}$ is said to be an even function (mod r) or an r-even function if $f(\gcd(n,r)) = f(n)$ for all $n \in \mathbb{N}$, where $r \in \mathbb{N}$ is fixed. Identity (1.3.1) shows that for fixed r, the function $n \mapsto c_r(n)$ is an even function (mod r).

The set $\mathcal{E}_r$ of r-even functions forms a $d(r)$ dimensional subspace of $\mathcal{P}_r$, where $d(r)$ is the number of positive divisors of r. A function $f \in \mathcal{A}$ is called even if

$$f \in \bigcup_{r \in \mathbb{N}} \mathcal{E}_r.$$

The functions g_d with $d \mid r$ given by

$$g_d(n) = \begin{cases} 1 & \text{if } \gcd(n,r) = d, \\ 0 & \text{if } \gcd(n,r) \neq d \end{cases}$$

form a basis of $\mathcal{E}_r$. This basis can be replaced by the following one. The Ramanujan sums c_q with $q \mid r$ form a basis of the subspace $\mathcal{E}_r$.

Consequently, every r-even function f has a (Ramanujan-)Fourier expansion of the form

$$f(n) = \sum_{d \mid r} h(d) c_d(n) \qquad (n \in \mathbb{N}), \tag{1.6.7}$$

where the (Ramanujan-)Fourier coefficients $h(d)$ are uniquely determined and are given by

$$h(d) = \frac{1}{r} \sum_{e \mid r} f(e) c_{r/e}(r/d) \qquad (d \mid r), \tag{1.6.8}$$

and the function h is also r-even. Notation:

$$h(d) = \alpha_f(d) \quad (d \mid r).$$

Note that $(\mathcal{E}_r, \otimes)$ is a subsemigroup of $(\mathcal{P}_r, \otimes)$ and

$$\alpha_{f \otimes h}(d) = r \alpha_f(d) \alpha_h(d) \quad (d \mid r).$$

The following result is a characterization of r-even functions. Given $r \in \mathbb{N}$ let

$$\mathcal{E}'_r = \{ f \in \mathcal{A} : f(n) = 0 \text{ for every } n \nmid r \}.$$

Proposition 1.6.1 ([121, Proposition 1]). *Let $f \in \mathcal{A}$. The following assertions are equivalent*:

(i) $f \in \mathcal{E}_r$,

(ii) $f(n) = \sum_{d|\gcd(n,r)} (\mu * f)(d)$ $(n \in \mathbb{N})$,

(iii) $\mu * f \in \mathcal{E}'_r$.

Proof. Let $\mu * f \in \mathcal{E}'_r$. Then for every $n \in \mathbb{N}$,

$$f(n) = \sum_{d|n} (\mu * f)(d) = \sum_{d|n,\, d|r} (\mu * f)(d) = \sum_{d|\gcd(n,r)} (\mu * f)(d)$$

$$= (\mu * f * \mathbf{1})(\gcd(n,r)) = f(\gcd(n,r)),$$

where $\mathbf{1}$ is the constant 1 function. This shows the implications (iii) $\Rightarrow$ (ii) $\Rightarrow$ (i).

Now we show that (i) $\Rightarrow$ (iii). Assume that $f \in \mathcal{E}_r$ and $\mu * f \notin \mathcal{E}'_r$, i.e., $(\mu * f)(n) \neq 0$ for some $n \in \mathbb{N}$ with $n \nmid r$. Consider the minimal $n \in \mathbb{N}$ with this property. Then all proper divisors d of n with $(\mu * f)(d) \neq 0$ divide r so that

$$f(n) = \sum_{d|n} (\mu * f)(d) = \sum_{d|\gcd(n,r)} (\mu * f)(d) + (\mu * f)(n)$$

$$= f(\gcd(n,r)) + (\mu * f)(n) \neq f(\gcd(n,r)),$$

giving $f \notin \mathcal{E}_r$. $\qquad\qquad\square$

1.6.3. The DFT of even functions (mod r)

Observe that the Ramanujan sum $c_r(\cdot)$ is the DFT of the function ϱ_r defined by

$$\varrho_r(n) = \begin{cases} 1 & \text{if } (n,r) = 1, \\ 0 & \text{otherwise,} \end{cases} \qquad (1.6.9)$$

i.e., the principal character (mod r). Thus, for r fixed,

$$\widehat{\varrho_r} = c_r, \qquad \widehat{c_r} = r\varrho_r. \qquad (1.6.10)$$

The DFT of an r-even function is given by the following result.

Proposition 1.6.2 ([121, Proposition 2]). *For every $r \in \mathbb{N}$, the DFT is an automorphism of $\mathcal{E}_r$. For any $f \in \mathcal{E}_r$,*

$$\widehat{f}(n) = \sum_{d|r} f(d) c_{r/d}(n) \qquad (n \in \mathbb{N}) \tag{1.6.11}$$

and the IDFT is given by

$$f(n) = \frac{1}{r} \sum_{d|r} \widehat{f}(d) c_{r/d}(n) \qquad (n \in \mathbb{N}). \tag{1.6.12}$$

Proof.   By the definition of r-even functions and grouping the terms in (1.6.3) according to the values $d = \gcd(k, r)$, we obtain

$$\widehat{f}(n) = \sum_{d|r} f(d) \sum_{\substack{1 \le j \le r/d \\ \gcd(j, r/d) = 1}} e(-jn/(r/d))$$

$$= \sum_{d|r} f(d) c_{r/d}(n)$$

giving (1.6.11) and also that $\widehat{f} \in \mathcal{E}_r$. Now applying (1.6.11) for $\widehat{f}$ (instead of f) and using that $\widehat{\widehat{f}} = rf$ we have (1.6.12). □

Corollary 1.6.3 ([121, Corollary 3]). *Let f be an r-even function. Then*

$$\widehat{f}(n) = \sum_{d|\gcd(n,r)} d\,(\mu * f)(r/d) \qquad (n \in \mathbb{N}), \tag{1.6.13}$$

and

$$(\mu * \widehat{f})(n) = \begin{cases} n(\mu * f)(r/n) & \text{if } n \mid r, \\ 0 & \text{otherwise.} \end{cases}$$

Next we show that some known properties of r-even functions and of Ramanujan sums can be obtained in a simple manner via the DFT.

Corollary 1.6.4 ([121, Appl. 5]). *Let $T_r(n, k)$ denote the number of (incongruent) solutions (mod r) of the congruence*

$$x_1 + \cdots + x_k \equiv n \pmod{r}$$

with

$$\gcd(x_1, r) = \cdots = \gcd(x_k, r) = 1.$$

Then

$$T_r(n, k) = \frac{1}{r} \sum_{d|r} (c_r(r/d))^k c_d(n) \qquad (n \in \mathbb{N}). \tag{1.6.14}$$

Proof. It is immediate from the definitions that

$$T_r(., k) = \underbrace{\varrho_r \otimes \cdots \otimes \varrho_r}_{k}, \tag{1.6.15}$$

where the function ϱ_r is defined by (1.6.9).

Therefore

$$\widehat{T_r}(., k) = (\widehat{\varrho_r})^k = (c_r)^k.$$

Now the IDFT formula (1.6.12) gives at once formula (1.6.14), which goes back to the work of H. Rademacher (1925) and A. Brauer (1926) and has been recovered several times (see [75, Chapter 3; 100, p. 41; 104]). □

We give a new proof of the following inversion formula of Cohen [29, Theorem 3]:

Corollary 1.6.5 ([121, Appl. 6]). *Let f and g be r-even functions and assume that f is defined by*

$$f(n) = \sum_{d|r} g(d) c_d(n) \quad (n \in \mathbb{N}).$$

Then

$$g(m) = \frac{1}{r} \sum_{d|r} f(r/d) c_d(n), \quad m = r/\gcd(n, r), \qquad (n \in \mathbb{N}).$$

Proof. Consider the function $G(n) = g(r/\gcd(n, r))$ which is also r-even. By Proposition 1.6.2,

$$\widehat{G}(n) = \sum_{d|r} G(r/d) c_d(n) = \sum_{d|r} g(d) c_d(n) = f(n).$$

Hence

$$rg(m) = rG(n) = \widehat{\widehat{G}}(n) = \widehat{f}(n) = \sum_{d|r} f(r/d) c_d(n). \qquad \square$$

The next result follows by the Parseval formula (1.6.5) and by grouping the terms of the right-hand side according to the values $\gcd(n,r)$.

Corollary 1.6.6 ([121, Appl. 8]). *If f is any r-even function, then*

$$\sum_{n=1}^{r} |\widehat{f}(n)|^2 = r \sum_{d|r} |f(d)|^2 \phi(r/d). \qquad (1.6.16)$$

For $f = \varrho_r$ we obtain the familiar formula

$$\sum_{n=1}^{r} (c_r(n))^2 = r\phi(r) \quad (r \in \mathbb{N}).$$

1.6.4. Menon-type identities concerning Dirichlet characters

We have the following general result.

Theorem 1.6.7 ([117, Theorem 2.1]). *Let $n,d \in \mathbb{N}$, $r,s \in \mathbb{Z}$ such that $d \mid n$. Let f be an even function (mod n). Then*

$$\sum_{\substack{k=1 \\ (k,n)=1 \\ k\equiv r \,(\mathrm{mod}\, d)}}^{n} f(k-s) = \begin{cases} \dfrac{\phi(n)}{\phi(d)} \displaystyle\sum_{\substack{e|n \\ (e,s)=1 \\ (e,d)|r-s}} \dfrac{(\mu * f)(e)}{\phi(e)} \phi((e,d)) & \text{if } (r,d)=1, \\[6mm] 0 & \text{if } (r,d)>1. \end{cases}$$
$$(1.6.17)$$

Corollary 1.6.8 ([117, Corollary 2.2]). *Let $n,d \in \mathbb{N}$, $r,s \in \mathbb{Z}$ such that $d \mid n$. Then*

$$\sum_{\substack{k=1 \\ (k,n)=1 \\ k\equiv r \,(\mathrm{mod}\, d)}}^{n} (k-s,n) = \begin{cases} \dfrac{\phi(n)}{\phi(d)} \displaystyle\sum_{\substack{e|n \\ (e,s)=1 \\ (e,d)|r-s}} \phi((e,d)) & \text{if } (r,d)=1, \\[6mm] 0 & \text{if } (r,d)>1. \end{cases} \qquad (1.6.18)$$

If $d = 1$ and $s = 1$, then (1.6.18) reduces to Menon's identity (1.1.12). For the Ramanujan sums, we deduce the following result.

Corollary 1.6.9 ([117, Corollary 2.3]). *Let* $n, d \in \mathbb{N}$, $r, s \in \mathbb{Z}$ *such that* $d \mid n$. *Then*

$$\sum_{\substack{k=1 \\ (k,n)=1 \\ k \equiv r \,(\mathrm{mod}\, d)}}^{n} c_n(k-s) = \begin{cases} \dfrac{\phi(n)}{\phi(d)} \displaystyle\sum_{\substack{e|n \\ (e,s)=1 \\ (e,d)|r-s}} \dfrac{e\mu(n/e)}{\phi(e)} \phi((e,d)) & \text{if } (r,d)=1, \\[3em] 0 & \text{if } (r,d)>1. \end{cases}$$

$$(1.6.19)$$

If $d = 1$, then (1.6.19) gives the first identity of the known formulas

$$\sum_{\substack{k=1 \\ (k,n)=1}}^{n} c_n(k-s) = \phi(n) \sum_{\substack{e|n \\ (e,s)=1}} \frac{e\mu(n/e)}{\phi(e)} = \mu(n)c_n(s),$$

the second one being the Brauer–Rademacher identity (cf. (1.1.11)).

Theorem 1.6.10 ([117, Theorem 2.4]). *Let* χ *be a Dirichlet character* (mod n) *with conductor* n^* ($n, n^* \in \mathbb{N}$, $n^* \mid n$). *Let* f *be an even function* (mod n) *and let* $s \in \mathbb{Z}$. *Then*

$$\sum_{k=1}^{n} f(k-s)\chi(k) = \phi(n)\chi^*(s) \sum_{\substack{d|n/n^* \\ (d,s)=1}} \frac{(\mu * f)(dn^*)}{\phi(dn^*)},$$

where χ^* *is the primitive character* (mod n^*) *that induces* χ.

Corollary 1.6.11 ([117, Corollary 2.5]). *Let* χ *be a Dirichlet character* (mod n) *with conductor* n^* ($n, n^* \in \mathbb{N}$, $n^* \mid n$) *and let* $s \in \mathbb{Z}$. *Then*

$$\sum_{k=1}^{n} (k-s,n)\chi(k) = \phi(n)\chi^*(s) \sum_{\substack{d|n/n^* \\ (d,s)=1}} 1. \qquad (1.6.20)$$

If $s = 1$, then (1.6.20) reduces to the next identity due to Zhao and Cao [130]:

$$\sum_{k=1}^{n} (k-1,n)\chi(k) = \phi(n)d(n/n^*) \quad (n \in \mathbb{N}). \qquad (1.6.21)$$

If χ is the principal character (mod n), that is $n^* = 1$, then (1.6.20) gives

$$\sum_{\substack{k=1 \\ (k,n)=1}}^{n} (k-s,n) = \phi(n) \sum_{\substack{d|n \\ (d,s)=1}} 1, \qquad (1.6.22)$$

which is valid for any $s \in \mathbb{Z}$. If $(s,n) = 1$, then the right-hand side of (1.6.22) is $\phi(n)d(n)$, like in Menon's classical identity (1.1.12).

Corollary 1.6.12 ([117, Corollary 2.6]). *Let χ be a Dirichlet character* (mod n) *with conductor n^* $(n, n^* \in \mathbb{N}, n^* \mid n)$ and let $s \in \mathbb{Z}$. Then*

$$\sum_{k=1}^{n} c_n(k-s)\chi(k) = n^*\phi(n)\chi^*(s) \sum_{\substack{d|n/n^* \\ (d,s)=1}} \frac{d\mu(n/(dn^*))}{\phi(dn^*)}.$$

We remark that the sums in Theorem 1.6.10 and Corollaries 1.6.11 and 1.6.12 vanish provided that $(s, n^*) > 1$.

Theorem 1.6.13 ([117, Theorem 2.7]). *Let χ be a primitive Dirichlet character* (mod n), *where $n \in \mathbb{N}$. Let f be an even function* (mod n) *and let $s \in \mathbb{Z}$. Then*

$$\sum_{k=1}^{n} f(k-s)\chi(k) = (\mu * f)(n)\chi(s).$$

Theorem 1.6.13 is a direct consequence of Theorem 1.6.10 by taking $d = n$. To give a short direct proof we need the following lemma (see, e.g., [79, Theorem 9.4]).

Lemma 1.6.14. *Let χ be a primitive character* (mod n). *Then for any $d \mid n$, $d < n$ and any $s \in \mathbb{Z}$,*

$$\sum_{\substack{k=1 \\ k \equiv s \,(\mathrm{mod}\ d)}}^{n} \chi(k) = 0.$$

Proof of Theorem 1.6.13. Note that, since f is an even function (mod n),

$$f(k) = f((k,n)) = \sum_{d|(k,n)} (\mu * f)(d). \tag{1.6.23}$$

By using (1.6.23) and Lemma 1.6.14, we have

$$\sum_{k=1}^{n} f(k - s)\chi(k) = \sum_{k=1}^{n} \chi(k) \sum_{e|(k-s,n)} (\mu * f)(e)$$

$$= \sum_{e|n} (\mu * f)(e) \sum_{\substack{k=1 \\ k \equiv s (\text{mod } e)}}^{n} \chi(k) = (\mu * f)(n)\chi(s),$$

which completes the proof. □

In particular, for $f(k) = c_n(k)$ we deduce the following result.

Corollary 1.6.15. *Let χ be a primitive Dirichlet character* (mod n), *where $n \in \mathbb{N}$. Let $s \in \mathbb{Z}$. Then*

$$\sum_{k=1}^{n} c_n(k - s)\chi(k) = n\,\chi(s).$$

1.6.5. Menon-type identities concerning additive characters

We recall the notation $n = \prod_p p^{\nu_p(n)}$ for the prime power factorization of $n \in \mathbb{N}$, the product being over the primes p, where all but a finite number of the exponents $\nu_p(n)$ are zero.

Consider the sum

$$S_f(n, k, s) = \sum_{\substack{a=1 \\ (a,n)=1}}^{n} f(a - s)\,e(ak/n), \tag{1.6.24}$$

where $f = f_n$ is an even function (mod n), $n \in \mathbb{N}$, $s, k \in \mathbb{Z}$.

If $f(a) = (a, n)$, then we have the sums investigated by Li and Kim [57]. They proved that for every $n \in \mathbb{N}$, $k \in \mathbb{Z}$ such that $\nu_p(n) - \nu_p(k) \neq 1$ for every prime $p \mid n$ one has (see [57, Theorem 3.3])

$$S^*(n, k) = \sum_{\substack{a=1 \\ (a,n)=1}}^{n} (a - 1, n)\,e(ak/n) = e(k/n)\phi(n)d((n,k)), \tag{1.6.25}$$

which reduces to (1.1.12) in the case $n \mid k$. Furthermore, Li and Kim [57, Theorem 3.5] established a formula for $S^*(n,k)$, which is valid for every $n \in \mathbb{N}$ and $k \in \mathbb{Z}$.

If f is the constant 1 function, then (1.6.24) reduces to the Ramanujan sum $c_n(k)$. The sum (1.6.24) can be viewed as the DFT of the function h_n attached to f_n and defined by $h_n(a) = f_n(a-s)$ if $(a,n)=1$ and $h_n(a)=0$ if $(a,n)>1$. However, if the function f_n is even (mod n), then h_n does not have this property, in general.

We have the following general results.

Theorem 1.6.16 ([120, Theorem 3.1]). *Let* $n \in \mathbb{N}$, $s,k \in \mathbb{Z}$ *and let* $f = f_n$ *be an even function* (mod n). *Then*

$$S_f(n,k,s) = n \sum_{\substack{d\mid n \\ (d,s)=1}} \frac{(\mu * f_n)(d)}{d} \sum_{\substack{\delta\mid n \\ (\delta,d)=1 \\ \frac{n}{d\delta}\mid k}} \frac{\mu(\delta)}{\delta} e(\delta\delta' k/n)$$

$$= \sum_{e\mid (n,k)} e \sum_{\substack{d\delta=n/e \\ (d,\delta s)=1}} (\mu * f_n)(d)\mu(\delta)e(\delta\delta' k/n),$$

where $\delta' \in \mathbb{Z}$ *is such that* $\delta\delta' \equiv s$ (mod d).

Theorem 1.6.17 ([120, Theorem 3.2]). *Under the assumptions of Theorem 1.6.16 and with* $\nu_p(n) \geq \nu_p(k) + 2$ *for every prime* $p \mid n$, *we have*

$$S_f(n,k,s) = e(ks/n) \sum_{\substack{d\mid (n,k) \\ (n/d,s)=1}} d\,(\mu * f_n)(n/d). \qquad (1.6.26)$$

Corollary 1.6.18 ([120, Corollary 3.4]). *Theorems 1.6.16 and 1.6.17 apply to the following functions:*

(i) $f_n(a) = (a,n)^m$ $(m \in \mathbb{R})$ *with* $(\mu * f_n)(d) = J_m(d)$ $(d \mid n)$;

(ii) $f_n(a) = \sigma_m((a,n)) := \sum_{d\mid(a,n)} d^m$ $(m \in \mathbb{R})$ *with* $(\mu * f_n)(d) = d^m$ $(d \mid n)$;

(iii) $f_n(a) = c_n(a)$ *with* $(\mu * f_n)(d) = d\mu(n/d)$ $(d \mid n)$.

The sums $S_f(n,k,s)$ defined by (1.6.24) have the following modified multiplicativity property.

Theorem 1.6.19 ([120, Theorem 3.5]). *Let* $(f_n)_{n\in\mathbb{N}}$ *be a sequence of even functions* (mod n) *such that* $n \mapsto f_n(a)$ *is multiplicative for every fixed*

$a \in \mathbb{Z}$. *Let* $n_1, n_2 \in \mathbb{N}$, $(n_1, n_2) = 1$, $s, k \in \mathbb{Z}$. *Then*

$$S_f(n_1 n_2, k, s) = S_f(n_1, k n_2', s) S_f(n_2, k n_1', s),$$

where $n_1', n_2' \in \mathbb{Z}$ *are such that*

$$n_1 n_1' \equiv 1 \pmod{n_2} \text{ and } n_2 n_2' \equiv 1 \pmod{n_1}.$$

Note that under assumptions of Theorem 1.6.19, $f_n(a)$ is multiplicative viewed as a function of two variables (see [121, Proposition 4]). Examples of sequences of functions satisfying the assumptions of Theorem 1.6.19 are the sequence of the Ramanujan sums $(c_n(\cdot))_{n \in \mathbb{N}}$ and $(f_n)_{n \in \mathbb{N}}$, where

$$f_n(a) = F((a, n))$$

and F is an arbitrary multiplicative function.

Theorem 1.6.20 ([120, Theorem 3.6]). *Under the assumptions of Theorem 1.6.19 and if* $\nu_p(n) - \nu_p(k) \neq 1$ *for every prime* $p \mid n$, *we have*

$$S_f(n, k, s) = e(ks/n)\phi(n_2) \sum_{\substack{d \mid (n_1, k) \\ (n_1/d, s) = 1}} d\,(\mu * f_{n_1})(n_1/d) \sum_{\substack{d \mid n_2 \\ (d, s) = 1}} \frac{(\mu * f_{n_2})(d)}{\phi(d)},$$

where $n = n_1 n_2$ *such that*

$$\nu_p(n) \geq \nu_p(k) + 2 \text{ for every prime } p \mid n_1,$$

and

$$\nu_p(n) \leq \nu_p(k) \text{ for every prime } p \mid n_2.$$

In the special case $f_n(a) = c_n(a)$ we deduce the following identity.

Corollary 1.6.21 ([120, Corollary 3.8(iii)]). *Let* $n \in \mathbb{N}$, $s, k \in \mathbb{Z}$ *such that* $(s, n) = 1$ *and* $\nu_p(n) - \nu_p(k) \neq 1$ *for every prime* $p \mid n$. *Let* n_1 *and* n_2 *be defined as in Theorem 1.6.20. Then*

$$\sum_{\substack{a=1 \\ (a,n)=1}}^{n} c_n(a-s)e(ak/n) = \begin{cases} e(ks/n)n_1 & \text{if } (n_1, k) = 1 \text{ and } n_2 \text{ is squarefree,} \\ 0 & \text{otherwise.} \end{cases}$$

$$(1.6.27)$$

For the proofs of the results of Section 1.6.5, Tóth [120] considered the sum

$$T_n(k, s, d) = \sum_{\substack{a=1 \\ (a,n)=1 \\ a \equiv s \,(\mathrm{mod}\ d)}}^{n} e(ak/n), \tag{1.6.28}$$

which reduces to Ramanujan's sum $c_n(k)$ if $d = 1$.

Lemma 1.6.22 ([120, Lemma 4.1]). *Let* $n, d \in \mathbb{N}$, $k, s \in \mathbb{Z}$ *such that* $d \mid n$. *Then*

$$T_n(k, s, d) = \begin{cases} \dfrac{n}{d} \displaystyle\sum_{\substack{\delta \mid n \\ (\delta,d)=1 \\ \frac{n}{d\delta} \mid k}} \dfrac{\mu(\delta)}{\delta} e(\delta\delta' k/n) & \text{if } (s, d) = 1, \\[4ex] 0 & \text{otherwise,} \end{cases}$$

where $\delta' \in \mathbb{Z}$ *such that* $\delta\delta' \equiv s$ (mod d).

Note that in the case $d = 1$, this recovers the familiar formula (1.3.1).

1.7. Counting Solutions of Congruences in Several Variables

1.7.1. Linear congruences with constraints

A direct generalization of the interpretation of the orbicyclic function E defined in the Introduction is the following. Let $M \in \mathbb{N}$ and let $\mathcal{D}_k(M)$ ($1 \leq k \leq r$) be arbitrary nonempty subsets of the set of (positive) divisors of M. For an integer n, let $N_n(M, \mathcal{D}_1, \ldots, \mathcal{D}_r)$ denote for the number of incongruent solutions $(x_1, \ldots, x_r) \in \mathbb{Z}^r$ of the congruence

$$x_1 + \cdots + x_r \equiv n \ (\mathrm{mod}\ M), \tag{1.7.1}$$

satisfying $\gcd(x_1, M) \in \mathcal{D}_1, \ldots, \gcd(x_r, M) \in \mathcal{D}_r$.

The orbicyclic function E is recovered in the case

$$\mathcal{D}_k = \{M/m_k\} \quad (1 \leq k \leq r) \text{ and } n = 0.$$

Some other special cases of the function $N_n(M, \mathcal{D}_1, \ldots, \mathcal{D}_r)$ were investigated earlier by several authors. The case

$$\mathcal{D}_k = \{1\}, \text{ i.e., } \gcd(x_k, M) = 1 \ (1 \leq k \leq r)$$

was considered for the first time by Rademacher [90] in 1925 and Brauer [22] in 1926. It was recovered by Nicol and Vandiver [87] in 1954, Cohen [28] in 1955, Rearick [96] in 1963, and others. The case $r = 2$ was treated by Alder [1] in 1958. Also see Corollary 1.6.4, using properties of the DFT for the proof, and the paper by K. Bibak, B. M. Kapron, V. Srinivasan, R. Tauraso, and L. Tóth [20].

The general case of arbitrary subsets $\mathcal{D}_k$ was investigated, among others, by McCarthy [74] in 1975 and by Spilker [104] in 1996. One has

$$N_n(M, \mathcal{D}_1, \ldots, \mathcal{D}_r) = \frac{1}{M} \sum_{d \mid M} c_{M/d}(n) \prod_{i=1}^{r} \sum_{e \in \mathcal{D}_i(M)} c_{M/e}(d). \qquad (1.7.2)$$

The proof of formula (1.7.2) given in [104, Section 4], see also [75, Chapter 3], uses properties of r-even functions, Cauchy products and Ramanujan–Fourier expansions of functions.

In what follows, we give a simple direct proof of (1.7.2) in the case

$$\mathcal{D}_k(M) = \{d_k\} \quad (1 \leq k \leq r)$$

(see [113, Section 6]).

Proposition 1.7.1. *Let* $M \in \mathbb{N}$, $n \in \mathbb{Z}$ *and let* $d_1, \ldots, d_r \mid M$. *Then*

$$N_n(M, \{d_1\}, \ldots, \{d_r\}) = \frac{1}{M} \sum_{k=1}^{M} c_{M/d_1}(k) \cdots c_{M/d_r}(k) e(-kn/M) \qquad (1.7.3)$$

$$= \frac{1}{M} \sum_{\delta \mid M} c_{M/d_1}(\delta) \cdots c_{M/d_r}(\delta) c_{M/\delta}(n). \qquad (1.7.4)$$

Proof.   Only the identity

$$\sum_{k=1}^{M} e(kn/M) = \begin{cases} M, & M \mid n, \\ 0, & M \nmid n, \end{cases} \tag{1.7.5}$$

and the definition of Ramanujan sums are required. By the definition of $N_n(M, \{d_1\}, \ldots, \{d_r\})$, we have

$$N := N_n(M, \{d_1\}, \ldots, \{d_r\})$$

$$= \frac{1}{M} \sum_{\substack{1 \le x_1 \le M \\ \gcd(x_1, M) = d_1}} \cdots \sum_{\substack{1 \le x_r \le M \\ \gcd(x_1, M) = d_r}} \sum_{k=1}^{M} e(k(x_1 + \cdots + x_r - n)/M)$$

$$= \frac{1}{M} \sum_{k=1}^{M} e(-kn/M) \sum_{\substack{1 \le x_1 \le M \\ \gcd(x_1, M) = d_1}} e(kx_1/M) \cdots \sum_{\substack{1 \le x_r \le M \\ \gcd(x_r, M) = d_r}} e(kx_r/M),$$

and denoting $x_k = d_k y_k$, $\gcd(y_k, M/d_k) = 1$ $(1 \le k \le M)$,

$$N = \frac{1}{M} \sum_{k=1}^{M} e(-kn/M) \sum_{\substack{1 \le y_1 \le M/d_1 \\ \gcd(y_1, M/d_1) = 1}} e(ky_1/(M/d_1))$$

$$\cdots \sum_{\substack{1 \le y_r \le M/d_r \\ \gcd(y_r, M/d_r) = 1}} e(ky_r/(M/d_r))$$

$$= \frac{1}{M} \sum_{k=1}^{M} e(-kn/M) c_{M/d_1}(k) \cdots c_{M/d_r}(k).$$

This proves (1.7.3). To obtain (1.7.4), we use $c_{M/d_i}(k) = c_{M/d_i}(\gcd(k, M))$ and group the terms according to the values of $\gcd(k, M) = \delta$.   □

In the case

$$M = m, \quad \mathcal{D}_k = \{m/m_k\} \ (1 \le k \le r) \text{ and } n = 0,$$

(1.7.2) reduces to the following result concerning the function $E(m_1, \ldots, m_r)$.

Proposition 1.7.2 ([113, Proposition 9]). *For any* $m_1, \ldots, m_r \in \mathbb{N}$,

$$E(m_1, \ldots, m_r) = \frac{1}{m} \sum_{d \mid m} c_{m_1}(d) \cdots c_{m_r}(d)\phi(m/d). \qquad (1.7.6)$$

1.7.2. Quadratic congruences

Consider the quadratic congruence

$$a_1 x_1^2 + \cdots + a_k x_k^2 \equiv n \pmod{r}, \qquad (1.7.7)$$

where $n \in \mathbb{Z}$, $\mathbf{a} = (a_1, \ldots, a_k) \in \mathbb{Z}^k$ and let $N_k(n, r, \mathbf{a})$ denote the number of its incongruent solutions $(x_1, \ldots, x_k) \in \mathbb{Z}^k$.

We evaluate $N_k(n, r, \mathbf{a})$ using the quadratic Gauss sum $S(\ell, r)$ defined by

$$S(\ell, r) = \sum_{j=1}^{r} e(\ell j^2/r) \quad (\ell, r \in \mathbb{N}, \gcd(\ell, r) = 1). \qquad (1.7.8)$$

We have the following general results.

Proposition 1.7.3 ([114, Proposition 1]). *For every* $k, r \in \mathbb{N}$, $n \in \mathbb{Z}$, $\mathbf{a} = (a_1, \ldots, a_k) \in \mathbb{Z}^k$ *we have*

$$N_k(n, r, \mathbf{a}) = r^{k-1} \sum_{d \mid r} \frac{1}{d^k} \sum_{\substack{\ell=1 \\ (\ell, d) = 1}}^{d} e(-\ell n/d) S(\ell a_1, d) \cdots S(\ell a_k, d).$$

Using the known result that for every r odd and $\ell \in \mathbb{N}$ such that $\gcd(\ell, r) = 1$,

$$S(\ell, r) = \begin{cases} \left(\dfrac{\ell}{r}\right)\sqrt{r}, & r \equiv 1 \pmod{4}, \\[2mm] i\left(\dfrac{\ell}{r}\right)\sqrt{r}, & r \equiv -1 \pmod{4}, \end{cases} \qquad (1.7.9)$$

where $\left(\frac{\ell}{r}\right)$ is the Jacobi symbol (see [42, Theorem 7.5.6]) we obtain the next identity.

Proposition 1.7.4 ([114, Proposition 2]). *Assume that* $k, r \in \mathbb{N}$, r *is odd,* $n \in \mathbb{Z}$ *and* $\mathbf{a} = (a_1, \ldots, a_k) \in \mathbb{Z}^k$ *is such that* $\gcd(a_1 \cdots a_k, r) = 1$.

Then

$$N_k(n, r, \mathbf{a}) = r^{k-1} \sum_{d|r} \frac{i^{k(d-1)^2/4}}{d^{k/2}} \left(\frac{a_1 \cdots a_k}{d}\right) \sum_{\substack{\ell=1 \\ (\ell,d)=1}}^{d} \left(\frac{\ell}{d}\right)^k e(-\ell n/d).$$

$$(1.7.10)$$

Furthermore, applying that for every ℓ odd,

$$S(\ell, 2^\nu) = \begin{cases} 0, & \nu = 1, \\ (1 + i^\ell)2^{\nu/2}, & \nu \text{ even}, \\ 2^{(\nu+1)/2} e(\ell/8), & \nu > 1 \text{ odd}, \end{cases}$$

we obtain the following proposition.

Proposition 1.7.5 ([114, Proposition 3]). *Assume that $k \in \mathbb{N}$, $r = 2^\nu$ ($\nu \in \mathbb{N}$), $n \in \mathbb{Z}$ and $\mathbf{a} = (a_1, \ldots, a_k) \in \mathbb{Z}^k$ such that $a_1, \ldots, a_k$ are odd. Then*

$$N_k(n, 2^\nu, \mathbf{a}) = 2^{\nu(k-1)} \left(1 + \sum_{t=1}^{\lfloor \nu/2 \rfloor} \frac{1}{2^{kt}} \sum_{\substack{\ell=1 \\ \ell \text{ odd}}}^{2^{2t}} e(-\ell n/2^{2t})(1 + i^{\ell a_1}) \cdots (1 + i^{\ell a_k}) \right.$$

$$\left. + \sum_{t=1}^{\lfloor (\nu-1)/2 \rfloor} \frac{1}{2^{kt}} \sum_{\substack{\ell=1 \\ \ell \text{ odd}}}^{2^{2t+1}} e(-\ell n/2^{2t+1} + \ell(a_1 + \cdots + a_k)/8) \right).$$

Now assume that k is even and r is odd. In this case, as a direct consequence of Proposition 1.7.4, we deduce for $N_k(n, r, \mathbf{a})$ the next formula in terms of the Ramanujan sums.

Proposition 1.7.6 ([27, Theorem 11 and Eq. (5.2)]; [114, Proposition 4]). *Assume that $k = 2m$ ($m \in \mathbb{N}$), $r \in \mathbb{N}$ is odd, $n \in \mathbb{Z}$, $\mathbf{a} = (a_1, \ldots, a_k) \in \mathbb{Z}^k$, $\gcd(a_1 \cdots a_k, r) = 1$. Then*

$$N_{2m}(n, r, \mathbf{a}) = r^{2m-1} \sum_{d|r} \frac{c_d(n)}{d^m} \left(\frac{(-1)^m a_1 \cdots a_{2m}}{d}\right).$$

In particular, if $a_1 \cdots a_r = 1$, then we obtain identities (1.1.9) and (1.1.10) given in the introduction. Note that in the cases $k = 2$ and $k = 4$, the following simple formulas are valid: for every r odd,

$$N_2(0,r) = r \sum_{d|r} (-1)^{(d-1)/2} \frac{\phi(d)}{d}, \qquad (1.7.11)$$

$$N_4(1,r) = r^3 \sum_{d|r} \frac{\mu(d)}{d^2} = r^3 \prod_{p|r} \left(1 - \frac{1}{p^2}\right). \qquad (1.7.12)$$

Remark 1.7.7. In the case k even and $a_1 = \ldots = a_k = 1$ for the proof of Proposition 1.7.6 it is sufficient to use the formula

$$S^2(\ell, r) = (-1)^{(r-1)/2} \quad (r \text{ odd}, \gcd(\ell, r) = 1)$$

instead of the much deeper result (1.7.9) giving the precise value of $S(\ell, r)$.

If k and r are odd, then we have the following proposition.

Proposition 1.7.8 ([27, Corollary 2]; [114, Proposition 14]). *Assume that $k = 2m + 1$ $(m \geq 0)$, $r \in \mathbb{N}$ is odd, $n \in \mathbb{Z}$ such that*

$$\gcd(n, r) = 1, \quad \mathbf{a} = (a_1, \ldots, a_k) \in \mathbb{Z}^k, \quad \gcd(a_1 \cdots a_k, r) = 1.$$

Then

$$N_{2m+1}(n, r, \mathbf{a}) = r^{2m} \sum_{d|r} \frac{\mu^2(d)}{d^m} \left(\frac{(-1)^m n a_1 \cdots a_{2m+1}}{d}\right).$$

Proof. Consider the sum

$$T(n, r) = \sum_{\substack{j=1 \\ \gcd(j,r)=1}}^{r} \left(\frac{j}{r}\right) e(jn/r).$$

For r odd, the Jacobi symbol $j \mapsto \left(\frac{j}{r}\right)$ is a real character (mod r) and

$$T(n, r) = \left(\frac{n}{r}\right) T(1, r)$$

holds if $\gcd(n, r) = 1$ (see, e.g., [42, Chapter 7]).

(i) If r is squarefree, then $j \mapsto \left(\frac{j}{r}\right)$ is a primitive character (mod r). Thus,

$$T(1, r) = \sqrt{r} \text{ for } \left(\frac{-1}{r}\right) = 1 \text{ and } T(1, r) = i\sqrt{r} \text{ for } \left(\frac{-1}{r}\right) = -1$$

(see [42, Theorem 7.5.8]), that is

$$
T(n,r) = \begin{cases} \left(\dfrac{n}{r}\right)\sqrt{r} & \text{if } r \equiv 1 \ (\mathrm{mod}\ 4), \\[2mm] i\left(\dfrac{n}{r}\right)\sqrt{r} & \text{if } r \equiv -1 \ (\mathrm{mod}\ 4). \end{cases} \tag{1.7.13}
$$

(ii) We show that if r is not squarefree, then $T(1,r) = 0$. Here r can be written as $r = p^2 s$, where p is a prime and by putting $j = ks + q$,

$$
T(1,r) = \sum_{q=1}^{s} \sum_{k=0}^{p^2-1} \left(\frac{ks+q}{r}\right) e((ks+q)/r)
$$

$$
= \sum_{q=1}^{s} \left(\frac{q}{s}\right) e(q/(p^2 s)) \sum_{k=0}^{p^2-1} e(k/p^2) = 0,
$$

since the inner sum is zero.

By applying Proposition 1.7.4 we deduce Proposition 1.7.8.   □

Proposition 1.7.9 ([27, Corollary 1; 114, Proposition 19]). *Assume that $k, r \in \mathbb{N}$ are odd, $n = 0$ and $\mathbf{a} = (a_1, \ldots, a_k) \in \mathbb{Z}^k$, $\gcd(a_1 \cdots a_k, r) = 1$. Then*

$$
N_k(0, r, \mathbf{a}) = r^{k-1} \sum_{d^2 \mid r} \frac{\phi(d)}{d^{k-1}},
$$

which does not depend on $\mathbf{a}$.

Proof.   From Proposition 1.7.4, we have

$$
N_k(0, r, \mathbf{a}) = r^{k-1} \sum_{d \mid r} \frac{i^{k(d-1)^2/4}}{d^{k/2}} \left(\frac{a_1 \cdots a_k}{d}\right) V(d),
$$

where for r odd, $V(r)$ is given by

$$
V(r) := T(0, r) = \sum_{\substack{j=1 \\ \gcd(j,r)=1}}^{r} \left(\frac{j}{r}\right),
$$

and we need to evaluate this sum.

If $r = t^2$ is a square, then

$$
\left(\frac{j}{r}\right) = \left(\frac{j}{t^2}\right) = 1
$$

for every j with $\gcd(j, r) = 1$ and deduce that $V(r) = \phi(r)$.

Now assume that r is not a square. Then, since r is odd, there is a prime $p > 2$ such that $r = p^\nu s$, where ν is odd and $\gcd(p, s) = 1$. Let c be a quadratic nonresidue (mod p) and consider the simultaneous congruences

$$x \equiv c \ (\text{mod } p), \quad x \equiv 1 \ (\text{mod } s).$$

By the Chinese Remainder Theorem, there exists a solution $x = j_0$ satisfying $\left(\frac{j_0}{r}\right) = -1$. Hence

$$V(r) = \sum_{\substack{j=1 \\ \gcd(j,r)=1}}^{r} \left(\frac{j j_0}{r}\right) = - \sum_{\substack{j=1 \\ \gcd(j,r)=1}}^{r} \left(\frac{j}{r}\right) = -V(r),$$

giving that $V(r) = 0$. We deduce

$$N_k(0, r, \mathbf{a}) = r^{k-1} \sum_{d^2 | r} \frac{i^{k(d^2-1)^2/4}}{d^k} \left(\frac{a_1 \cdots a_k}{d^2}\right) \phi(d^2) = r^{k-1} \sum_{d^2 | r} \frac{\phi(d^2)}{d^k}. \qquad \square$$

Now assume that $r = 2^\nu$ and $a_1 = \cdots = a_k = 1$. We have the following results.

Proposition 1.7.10 ([114, Proposition 21]). *If $k \in \mathbb{N}$ is even and $n \in \mathbb{Z}$ is odd, then $N_k(n, 2) = 2^{k-1}$ and for every $\nu \in \mathbb{N}$, $\nu \geq 2$,*

$$N_k(n, 2^\nu) = 2^{\nu(k-1)} \left(1 - \frac{1}{2^{k/2-1}} \cos\left(\frac{k\pi}{4} + \frac{n\pi}{2}\right)\right).$$

Proposition 1.7.11 ([114, Proposition 24]). *If $k = 4m$ ($m \in \mathbb{N}$) and $n = 0$, then, for every $\nu \in \mathbb{N}$,*

$$N_{4m}(0, 2^\nu) = 2^{\nu(4m-1)} \left(1 + \frac{(-1)^m (2^{(\nu-1)(2m-1)} - 1)}{2^{(\nu-1)(2m-1)}(2^{2m-1} - 1)}\right).$$

Proposition 1.7.12 ([114, Proposition 25]). *If $k = 4m + 2$ ($m \geq 0$) and $n = 0$, then, for every $\nu \in \mathbb{N}$,*

$$N_{4m+2}(0, 2^\nu) = 2^{\nu(4m+1)}.$$

Proposition 1.7.13 ([114, Proposition 26]). *If $k \in \mathbb{N}$ is odd and $n = 0$, then for every $\nu \in \mathbb{N}$,*

$$N_k(0, 2^\nu) = 2^{\nu(k-1)} \left(1 + \frac{(-1)^{(k^2-1)/8} \cdot (2^{(k-2)\lfloor \nu/2 \rfloor} - 1)}{2^{(k-2)\lfloor \nu/2 \rfloor - (k-3)/2}(2^{k-2} - 1)}\right).$$

Next we will point out the asymptotic formulas for

- $N_1(0, r)$, which can be considered as an analog of Dirichlet's formula for the divisor function $d(n)$,
- $N_1(1, r)$, related to the squarefree divisor problem, and
- $N_2(0, r)$, related to the Gauss circle problem.

Theorem 1.7.14 ([114, Proposition 28]). *We have*

$$\sum_{r \leq x} N_1(0, r) = \frac{3}{\pi^2} x \log x + cx + O(x^{2/3}),$$

where

$$c = \frac{3}{\pi^2} \left(3\gamma - 1 - \frac{2\zeta'(2)}{\zeta(2)} \right),$$

and $\gamma \doteq 0.577215$ is Euler's constant.

Proof. Consider the congruence $x^2 \equiv 0 \pmod{r}$. It follows from general results of paper [114], and can be deduced also directly, that for the number $N_1(0, r)$ of its solutions one has $N_1(0, p^\nu) = p^{\lfloor \nu/2 \rfloor}$ for every prime power p^ν ($\nu \geq 1$). This leads to the Dirichlet series representation

$$\sum_{r=1}^{\infty} \frac{N_1(0, r)}{r^s} = \frac{\zeta(2s-1)\zeta(s)}{\zeta(2s)}. \tag{1.7.14}$$

By the identity (1.7.14) we infer that for every $r \in \mathbb{N}$,

$$N_1(0, r) = \sum_{a^2 b^2 c = r} \mu(a) b.$$

Now using Dirichlet's hyperbola method we have

$$E(x) := \sum_{b^2 c \leq x} b = \sum_{b \leq x^{1/3}} b \sum_{c \leq x/b^2} 1 + \sum_{c \leq x^{1/3}} \sum_{b \leq (x/c)^{1/2}} b - \sum_{b \leq x^{1/3}} b \sum_{c \leq x^{1/3}} 1,$$

which gives by the trivial estimate (i.e., $|x - \lfloor x \rfloor| < 1$),

$$E(x) = \frac{1}{2} x \log x + \frac{1}{2}(3\gamma - 1)x + O(x^{2/3}).$$

Now,

$$\sum_{r \leq x} N_1(0, r) = \sum_{a \leq x^{1/2}} \mu(a) E(x/a^2)$$

and elementary computations complete the proof. □

Theorem 1.7.15 ([114, Proposition 30]).

$$\sum_{r \le x} N_1(1,r) = \frac{6}{\pi^2} x \log x + c_1 x + O(x^{1/2} \delta(x)),$$

where

$$c_1 = \frac{6}{\pi^2} \left(2\gamma - 1 - \frac{\log 2}{2} - \frac{2\zeta'(2)}{\zeta(2)} \right)$$

and

$$\delta(x) = \exp(-c(\log x)^{3/5}(\log\log x)^{-1/5}), \qquad (1.7.15)$$

c being a positive constant. If the Riemann Hypothesis (RH) is true, then the error term is $O(x^{4/11+\varepsilon})$ for every $\varepsilon > 0$.

Proof. It is known that for the number $N_1(1,r)$ of solutions of the congruence

$$x^2 \equiv 1 \pmod{r}$$

one has

$$N_1(1,p^\nu) = 2 \quad \text{for every prime } p > 2 \text{ and every } \nu \ge 1,$$

$$N_1(1,2) = 1, \ N_1(1,4) = 2, \ N_1(1,2^\nu) = 4 \quad \text{for every } \nu \ge 3$$

(see [103, sequence A060594]). The Dirichlet series representation

$$\sum_{r=1}^{\infty} \frac{N_1(1,r)}{r^s} = \frac{\zeta^2(s)}{\zeta(2s)} \left(1 - \frac{1}{2^s} + \frac{2}{2^{2s}} \right) \qquad (1.7.16)$$

shows that estimating the sum $\sum_{r \le x} N_1(1,r)$ is closely related to the squarefree divisor problem concerning the function $d^{(2)}(n) = 2^{\omega(n)}$. Here

$$\sum_{n=1}^{\infty} \frac{d^{(2)}(n)}{n^s} = \frac{\zeta^2(s)}{\zeta(2s)}. \qquad (1.7.17)$$

By the identities (1.7.16) and (1.7.17) it follows that for every $r \in \mathbb{N}$,

$$N_1(1,r) = \sum_{ab=r} d^{(2)}(a)h(b),$$

where the multiplicative function h is defined by

$$h(p^\nu) = \begin{cases} -1, & p = 2, \nu = 1, \\ 2, & p = 2, \nu = 2, \\ 0 & \text{otherwise.} \end{cases}$$

The proof is completed by using the convolution method and the known result

$$\sum_{n \le x} d^{(2)}(n) = \frac{6}{\pi^2} x \left(\log x + 2\gamma - 1 - \frac{2\zeta'(2)}{\zeta(2)} \right) + O(R(x)), \qquad (1.7.18)$$

with

$$R(x) \ll x^{1/2} \delta(x).$$

(see [107]). If RH is true, then use

$$R(x) \ll x^{4/11+\varepsilon},$$

an estimate due to Baker [9].   □

Theorem 1.7.16 ([114, Proposition 34]).

$$\sum_{r \le x} N_2(0, r) = \frac{\pi}{8G} x^2 + O(x^{547/416} (\log x)^{26947/8320}),$$

where

$$G = \sum_{n=0}^{\infty} \frac{(-1)^n}{(2n+1)^2} \doteq 0.915956$$

is the Catalan constant.[a]

Proof.   Here $N_2(0, r)$ is the number of solutions of the congruence $x^2 + y^2 \equiv 0 \pmod{r}$. It is the sequence A086933 in [103], and for r odd it is given by identity (1.7.11). Furthermore,

$$N_2(0, 2^\nu) = 2^\nu \quad (\nu \ge 1) \quad (\text{cf. [114, Proposition 25]}).$$

---

[a] A better error term is $O(x^{2165/1648})$ by using a recent improvement by Bourgain and Watt [21] for the Gauss circle problem.

We deduce that for every prime power p^ν $(\nu \geq 1)$,

$$N_2(0, p^\nu) = \begin{cases} p^\nu(\nu + 1 - \nu/p), & p \equiv 1 \pmod 4, \nu \geq 1, \\ p^\nu, & p \equiv -1 \pmod 4, \nu \text{ even}, \\ p^{\nu-1}, & p \equiv -1 \pmod 4, \nu \text{ odd}, \\ 2^\nu, & p = 2, \nu \geq 1. \end{cases}$$

We obtain the Dirichlet series representation

$$\sum_{r=1}^{\infty} \frac{N_2(0, r)}{r^s} = \zeta(s-1) \prod_{p>2} \left(1 - \frac{(-1)^{(p-1)/2}}{p^s}\right) \left(1 - \frac{(-1)^{(p-1)/2}}{p^{s-1}}\right)^{-1}$$

$$= \zeta(s-1)L(s-1, \chi)L(s, \chi)^{-1},$$

where $L(s, \chi)$ is the Dirichlet series of χ, the nonprincipal character (mod 4). Therefore,

$$N_2(0, r) = (\mathrm{id} \cdot (\mathbf{1} * \chi) * \mu\chi)(r) \quad (r \in \mathbb{N}),$$

where $\mathbf{1}(n) = 1$, $\mathrm{id}(n) = n$. Here

$$4(\mathbf{1} * \chi)(n) = r_2(n)$$

is the number of ways n can be written as a sum of two squares. This shows that the sum

$$\sum_{r \leq x} N_2(0, r)$$

is closely related to the Gauss circle problem. According to the asymptotic formula due to Huxley [43],

$$\sum_{n \leq x} r_2(n) = \pi x + O(x^a(\log x)^b), \tag{1.7.19}$$

where $a = 131/416 \doteq 0.314903$ and $b = 26947/8320$. We deduce that

$$\sum_{r \leq x} N_2(0, r) = \frac{1}{4} \sum_{d \leq x} \mu(d)\chi(d) \sum_{n \leq x/d} nr_2(n).$$

Now partial summation on (1.7.19) and usual estimates give the result. □

Remark 1.7.17. See the paper by Finch, Martin, and Sebah [36] for asymptotic formulas on the number of solutions of the higher degree congruences

$$x^\ell \equiv 0 \pmod{n} \quad \text{and} \quad x^\ell \equiv 1 \pmod{n},$$

respectively, where $\ell \in \mathbb{N}$. The error terms of our Theorems 1.7.14 and 1.7.15 are better than those of [36] applied to $\ell = 2$.

1.8. Polynomials of Which Coefficients are Ramanujan Sums

Consider the polynomials

$$R_n(x) = \sum_{k=0}^{n-1} c_n(k) x^k.$$

By using identity (1.3.1) one has for any $n \geq 1$,

$$R_n(x) = (1 - x^n) \sum_{d|n} \frac{d\mu(n/d)}{1 - x^d}. \tag{1.8.1}$$

Let $\gamma(n) = \prod_{p|n} p$ denote the squarefree kernel of n.

Theorem 1.8.1 ([111, Theorem 3]). *Let $n \geq 1$.*

(i) *The number of nonzero coefficients of $R_n(x)$ is $\gamma(n)$.*

(ii) *The degree of $R_n(x)$ is $n - n/\gamma(n)$.*

(iii) *$R_n(x)$ has coefficients ± 1 if and only if n is squarefree and in this case the number of coefficients ± 1 of $R_n(x)$ is $\phi(n)$ for n odd and is $2\phi(n/2)$ for n even.*

Proof. For $n = 1$ the assertions hold true. Let $n = p_1^{a_1} \cdots p_r^{a_r} > 1$.

(i) We use that $c_n(k)$ is multiplicative in n and for any prime power p^a its values are given by (1.3.3).

Therefore, $c_n(k) \neq 0$ if and only if

$$p_1^{a_1-1} \mid k, \ldots, p_r^{a_r-1} \mid k,$$

i.e.,

$$k = p_1^{a_1-1} \cdots p_r^{a_r-1} m \quad \text{with } 0 \leq m < p_1 \cdots p_r = \gamma(n).$$

Hence the number of nonzero values of $c_n(k)$ is $\gamma(n)$.

(ii) By the proof of i) the largest k such that $c_n(k) \neq 0$ is

$$k = p_1^{a_1-1} \cdots p_r^{a_r-1}(p_1 \cdots p_r - 1) = n - n/\gamma(n),$$

and this is the degree of $R_n(x)$.

(iii) $c_n(k) = \pm 1$ if and only if

$$c_{p_i^{a_i}}(k) = \pm 1 \quad \text{for any } i \in \{1, \ldots, r\},$$

that is $a_i = 1$ for any i (n is squarefree) and either $p_i \nmid k$ or $p_i = 2 \mid k$ for any i.

Suppose that $n = p_1 \cdots p_r$ (squarefree). If n is odd, then by condition $p_i \nmid k$ for any i we have $(n, k) = 1$, hence the number of such values of k is $\phi(n)$. For n even, either $(k, n) = 1$ or $k = 2\ell$ with $(\ell, n/2) = 1$. We obtain that the number of such values of k is

$$\phi(n) + \phi(n/2) = 2\phi(n/2). \qquad \square$$

For any $n > 1$, one has

$$R_n(0) = c_n(0) = \phi(n) \quad \text{and} \quad R_n(1) = \sum_{k=0}^{n-1} c_n(k) = 0,$$

as it is well known. Hence $1 - x$ divides $R_n(x)$ for any $n > 1$. Now a look at the polynomials $R_n(x)$ suggests that $1 + x$ divides $R_n(x)$ for any $n > 2$ even. See Table 1.1 including the polynomials $R_n(x)$ for $1 \leq n \leq 20$. This is confirmed by the next result.

Theorem 1.8.2 ([111, Theorem 4]). *We have $R_2(-1) = 2$ and*

(i) *$R_n(-1) = \phi(n)$ for any $n \geq 1$ odd,*
(ii) *$R_n(-1) = 0$ for any $n > 2$ even,*
(iii) *the cyclotomic polynomial $\Phi_n(x)$ divides the polynomial $R_n(x) - n$ for any $n \geq 1$.*

If p is a prime, then it follows from (1.8.1) that

$$R_p(x) = (p - 1) - x - x^2 - \cdots - x^{p-1}. \qquad (1.8.2)$$

Table 1.1.   Polynomials $R_n(x)$ for $1 \le n \le 20$.

n	$R_n(x)$
1	1
2	$1 - x$
3	$2 - x - x^2 = (1 - x)(2 + x)$
4	$2 - 2x^2 = 2(1 - x)(1 + x)$
5	$4 - x - x^2 - x^3 - x^4 = (1 - x)(4 + 3x + 2x^2 + x^3)$
6	$2 + x - x^2 - 2x^3 - x^4 + x^5$ $= (1 - x)(2 - x)(1 + x)(1 + x + x^2)$
7	$6 - x - x^2 - x^3 - x^4 - x^5 - x^6$ $= (1 - x)(6 + 5x + 4x^2 + 3x^3 + 2x^4 + x^5)$
8	$4 - 4x^4 = 4(1 - x)(1 + x)(1 + x^2)$
9	$6 - 3x^3 - 3x^6 = 3(1 - x)(2 + x^3)(1 + x + x^2)$
10	$4 + x - x^2 + x^3 - x^4 - 4x^5 - x^6 + x^7 - x^8 + x^9$ $= (1 - x)(1 + x)(4 - 3x + 2x^2 - x^3)(1 + x + x^2 + x^3 + x^4)$
11	$10 - x - x^2 - x^3 - x^4 - x^5 - x^6 - x^7 - x^8 - x^9 - x^{10}$ $= (1 - x)(10 + 9x + 8x^2 + 7x^3 + 6x^4 + 5x^5 + 4x^6 + 3x^7 + 2x^8 + x^9)$
12	$4 + 2x^2 - 2x^4 - 4x^6 - 2x^8 + 2x^{10}$ $= 2(1 - x)(1 + x)(2 - x^2)(1 + x^2)(1 - x + x^2)(1 + x + x^2)$
13	$12 - x - x^2 - x^3 - x^4 - x^5 - x^6 - x^7 - x^8 - x^9 - x^{10} - x^{11} - x^{12}$ $= (1 - x)(12 + 11x + 10x^2 + 9x^3 + 8x^4 + 7x^5 + 6x^6 + 5x^7 + 4x^8$ $+ 3x^9 + 2x^{10} + x^{11})$
14	$6 + x - x^2 + x^3 - x^4 + x^5 - x^6 - 6x^7 - x^8 + x^9 - x^{10} + x^{11} - x^{12} + x^{13}$ $= (1 - x)(1 + x)(1 + x + x^2 + x^3 + x^4 + x^5 + x^6)(6 - 5x + 4x^2 - 3x^3$ $+ 2x^4 - x^5)$
15	$8 + x + x^2 - 2x^3 + x^4 - 4x^5 - 2x^6 + x^7 + x^8 - 2x^9 - 4x^{10} + x^{11}$ $\quad - 2x^{12} + x^{13} + x^{14}$ $= (1 - x)(1 + x + x^2 + x^3 + x^4)(8 - 7x + 5x^3 - 4x^4 + 3x^5 - x^7)(1 + x + x^2)$
16	$8 - 8x^8 = 8(1 - x)(1 + x)(1 + x^2)(1 + x^4)$
17	$16 - x - x^2 - x^3 - x^4 - x^5 - x^6 - x^7 - x^8 - x^9 - x^{10} - x^{11} - x^{12}$ $\quad - x^{13} - x^{14} - x^{15} - x^{16}$ $= (1 - x)(16 + 15x + 14x^2 + 13x^3 + 12x^4 + 11x^5 + 10x^6 + 9x^7 + 8x^8$ $+ 7x^9 + 6x^{10} + 5x^{11} + 4x^{12} + 3x^{13} + 2x^{14} + x^{15})$
18	$6 + 3x^3 - 3x^6 - 6x^9 - 3x^{12} + 3x^{15}$ $= 3(1 - x)(1 + x)(1 - x + x^2)(1 + x + x^2)(1 + x^3 + x^6)(2 - x^3)$
19	$18 - x - x^2 - x^3 - x^4 - x^5 - x^6 - x^7 - x^8 - x^9 - x^{10} - x^{11} - x^{12}$ $\quad - x^{13} - x^{14} - x^{15} - x^{16} - x^{17} - x^{18}$ $= (1 - x)(18 + 17x + 16x^2 + 15x^3 + 14x^4 + 13x^5 + 12x^6 + 11x^7 + 10x^8$ $+ 9x^9 + 8x^{10} + 7x^{11} + 6x^{12} + 5x^{13} + 4x^{14} + 3x^{15} + 2x^{16} + x^{17})$
20	$8 + 2x^2 - 2x^4 + 2x^6 - 2x^8 - 8x^{10} - 2x^{12} + 2x^{14} - 2x^{16} + 2x^{18}$ $= 2(1 - x)(1 + x)(1 + x^2)(4 - 3x^2 + 2x^4 - x^6)(1 + x + x^2 + x^3 + x^4)$ $\quad \times (1 - x + x^2 - x^3 + x^4)$

Also, if p, q are distinct primes, then

$$\begin{aligned}
R_{pq}(x) &= (p-1)(q-1) + x + x^2 + \cdots + x^{pq-1} \\
&\quad - p(x^p + x^{2p} + \cdots + x^{(q-1)p}) \\
&\quad - q(x^q + x^{2q} + \cdots + x^{(p-1)q}).
\end{aligned} \tag{1.8.3}$$

Next we show that the polynomials $R_n(x)$ have some properties which are similar to those of the cyclotomic polynomials $\Phi_n(x)$.

Theorem 1.8.3 ([111, Theorem 5]). (i) *If $n \geq 1$, then*

$$R_n(x) = \frac{n}{\gamma(n)} R_{\gamma(n)}(x^{n/\gamma(n)}). \tag{1.8.4}$$

(ii) *Let $n \geq 1$ and p be a prime. If $p \mid n$, then*

$$R_{np}(x) = pR_n(x^p).$$

If $p \nmid n$, then

$$R_{np}(x) = pR_n(x^p) - (1 + x^n + x^{2n} + \cdots + x^{(p-1)n})R_n(x). \tag{1.8.5}$$

(iii) *If $n > 1$, p is a prime and $p \nmid n$, then*

$$(1 - x^p) \mid R_{np}(x).$$

In particular, for any prime power p^k $(k \geq 1)$,

$$\begin{aligned}
R_{p^k}(x) &= p^{k-1} R_p(x^{p^{k-1}}) \\
&= p^{k-1}(p - 1 - x^{p^{k-1}} - x^{2p^{k-1}} - \cdots - x^{(p-1)p^{k-1}}) \tag{1.8.6}
\end{aligned}$$

and for $p = 2$,

$$R_{2^k}(x) = 2^{k-1}(1 - x^{2^{k-1}}). \tag{1.8.7}$$

Theorem 1.8.4 ([111, Theorem 6]). (i) *For any $n \geq 1$ odd,*

$$R_{2n}(x) = (1 - x^n)R_n(-x), \tag{1.8.8}$$

(ii) *More generally, for any $n \geq 1$ odd and any $k \geq 1$,*

$$R_{2^k n}(x) = 2^{k-1}(1 - x^{2^{k-1}n})R_n(-x^{2^{k-1}}). \tag{1.8.9}$$

Theorem 1.8.5 ([111, Theorem 7]). (i) *If* $n = p^k$, *p prime*, $k \geq 1$, *then*

$$(1 - x^{p^{k-1}}) \mid R_n(x). \tag{1.8.10}$$

(ii) *If* $n = 2^k m$, $k \geq 1$, $m > 1$ *odd, then*

$$(1 - x^{n/2})(1 + x^{n/\gamma(n)}) \mid R_n(x). \tag{1.8.11}$$

(iii) *If* $n = p^k m$, $p > 2$ *prime*, $k \geq 1$, $m > 1$ *odd*, $p \nmid m$, *then*

$$(1 - x^{pn/\gamma(n)}) \mid R_n(x). \tag{1.8.12}$$

(iv) *If* $n = 2^k m$, $k \geq 1$, $m > 1$ *odd, m has at least two prime divisors,*
p prime, $p \mid m$, *then*

$$(1 - x^{n/2})(1 + x^{pn/\gamma(n)}) \mid R_n(x). \tag{1.8.13}$$

As examples, Theorem 1.8.5 gives that

$$(1 - x^9)(1 + x^3) \mid R_{18}(x) \quad \text{and} \quad (1 - x^{15})(1 + x^3) \mid R_{30}(x).$$

It is possible to deduce from Theorem 1.8.5 other divisibility properties for the polynomials $R_n(x)$, e.g., the following one.

Theorem 1.8.6 ([111, Theorem 8]). *For any* $k \geq 1$ *and* $m > 1$,

$$(1 + x^{2^{k-1}}) \mid R_{2^k m}(x). \tag{1.8.14}$$

By using Hölder's formula (1.3.5), another representation of the polynomials $R_n(x)$ is obtained.

Theorem 1.8.7 ([111, Theorem 9]). *For any* $n \geq 1$,

$$R_n(x) = \phi(n) \left(1 - x^n + \sum_{d \mid n} \frac{\mu(d)}{\phi(d)} \Psi_d(x^{n/d}) \right), \tag{1.8.15}$$

where

$$\Psi_n(x) = \sum_{j \in A_n} x^j.$$

Table 1.1 was produced using software Maple. The polynomials $R_n(x)$ were generated by the following procedure:

```
with(numtheory): Ramanujanpol:= proc(n,x) local a, k: a:= 0:
for k from 0 to n-1 do a:=a+phi(n)*mobius(n/gcd(n,k))/phi(n/
gcd(n,k))*x^k: od: RETURN(R[n](x)=a) end;
```

Table 1.2. Polynomials $T_n(x)$ for $1 \le n \le 20$.

n	$T_n(x)$
1	1
2	$1 + x$
3	$2 + x + x^2$
4	$2 + 2x^2 = 2(1 + x^2)$
5	$4 + x + x^2 + x^3 + x^4$
6	$2 + x + x^2 + 2x^3 + x^4 + x^5$ $= (1 + x)(1 - x + x^2)(2 + x + x^2)$
7	$6 + x + x^2 + x^3 + x^4 + x^5 + x^6$
8	$4 + 4x^4 = 4(1 + x^4)$
9	$6 + 3x^3 + 3x^6$
10	$4 + x + x^2 + x^3 + x^4 + 4x^5 + x^6 + x^7 + x^8 + x^9$ $= (1 + x)(4 + x + x^2 + x^3 + x^4)(1 - x + x^2 - x^3 + x^4)$
11	$10 + x + x^2 + x^3 + x^4 + x^5 + x^6 + x^7 + x^8 + x^9 + x^{10}$
12	$4 + 2x^2 + 2x^4 + 4x^6 + 2x^8 + 2x^{10}$ $= 2(1 + x^2)(2 + x^2 + x^4)(1 - x^2 + x^4)$
13	$12 + x + x^2 + x^3 + x^4 + x^5 + x^6 + x^7 + x^8 + x^9 + x^{10} + x^{11} + x^{12}$
14	$6 + x + x^2 + x^3 + x^4 + x^5 + x^6 + 6x^7 + x^8 + x^9 + x^{10} + x^{11} + x^{12} + x^{13}$ $= (1 + x)(1 - x + x^2 - x^3 + x^4 - x^5 + x^6)(6 + x + x^2 + x^3 + x^4 + x^5 + x^6)$
15	$8 + x + x^2 + 2x^3 + x^4 + 4x^5 + 2x^6 + x^7 + x^8 + 2x^9 + 4x^{10} + x^{11}$ $+ 2x^{12} + x^{13} + x^{14}$
16	$8 + 8x^8 = 8(1 + x^8)$
17	$16 + x + x^2 + x^3 + x^4 + x^5 + x^6 + x^7 + x^8 + x^9 + x^{10} + x^{11} + x^{12}$ $+ x^{13} + x^{14} + x^{15} + x^{16}$
18	$6 + 3x^3 + 3x^6 + 6x^9 + 3x^{12} + 3x^{15}$ $= 3(1 + x)(1 - x + x^2)(2 + x^3 + x^6)(1 - x^3 + x^6)$
19	$18 + x + x^2 + x^3 + x^4 + x^5 + x^6 + x^7 + x^8 + x^9 + x^{10} + x^{11} + x^{12}$ $+ x^{13} + x^{14} + x^{15} + x^{16} + x^{17} + x^{18}$
20	$8 + 2x^2 + 2x^4 + 2x^6 + 2x^8 + 8x^{10} + 2x^{12} + 2x^{14} + 2x^{16} + 2x^{18}$ $= 2(1 + x^2)(4 + x^2 + x^4 + x^6 + x^8)(1 - x^2 + x^4 - x^6 + x^8)$

In paper [111], properties of the polynomials

$$T_n(x) = \sum_{k=0}^{n-1} |c_n(k)| x^k$$

were also considered and compared to those of the polynomials $R_n(x)$. See Table 1.2 including the polynomials $T_n(x)$ for $1 \le n \le 20$. For example, the next general results were proved.

Theorem 1.8.8 ([111, Theorem 11]). *For any $n \ge 1$,*

$$T_n(x) = (1 - x^n)\phi(n) \sum_{d|n} \frac{\mu^2(d) f_d(n/d)}{\phi(d)(1 - x^{n/d})},$$

where $f_k(n)$ denotes the function

$$f_k(n) = \prod_{\substack{p|n \\ p\nmid k}} \left(1 - \frac{1}{p-1}\right).$$

Theorem 1.8.9 ([111, Theorem 12]). *We have*

(i) $T_n(-1) = \phi(n)$ *for any $n \geq 1$ odd,*

(ii) $T_n(-1) = 0$ *for any $n = 4k + 2$, $k \geq 0$,*

(iii) $T_n(-1) = \phi(n)2^{\omega(n)}$ *for any $n = 4k$, $k \geq 1$,*

(iv) $T_n(\eta) = n\prod_{p|n}(1 - \frac{2}{p})$ *for any primitive nth root of unity η. The cyclotomic polynomial $\Phi_n(x)$ divides the polynomial $T_n(x)$ for any $n \geq 2$ even.*

Theorem 1.8.10 ([111, Theorem 13]). (i) *If $n \geq 1$, then*

$$T_n(x) = \frac{n}{\gamma(n)} T_{\gamma(n)}(x^{n/\gamma(n)}).$$

(ii) *Let $n \geq 1$ and p be a prime. If $p \mid n$, then*

$$T_{np}(x) = pT_n(x^p).$$

If $p \nmid n$, then

$$T_{np}(x) = (p-2)\phi(n)T_n(x^p) + (1 + x^n + x^{2n} + \cdots + x^{(p-1)n})T_n(x).$$

1.9. Analogs and Generalizations of Ramanujan Sums

1.9.1. Unitary Ramanujan sums

Unitary divisors and the unitary convolution of arithmetic functions were defined in Section 1.2. In the one variable case, the unitary convolution of the functions $f, g \in \mathcal{A}$ is

$$(f \times g)(n) = \sum_{d\|n} f(d)g(n/d). \tag{1.9.1}$$

The study of arithmetic functions of one and several variables defined by unitary divisors goes back to Vaidyanathaswamy [123] and Cohen [30]. The unitary analogs of the sum-of-divisors function σ and Euler's totient function ϕ are

$$\sigma^*(n) = \sum_{d||n} d,$$

respectively,

$$\phi^*(n) = \#\{j : 1 \leq j \leq n, (j,n)_* = 1\},$$

where

$$(j,n)_* = \max\{d : d \mid j, \, d \parallel n\}.$$

The unitary Möbius function μ^* is defined as the inverse of the constant 1 function with respect to the unitary convolution (1.9.1). One has

$$\mu^*(n) = (-1)^{\omega(n)}.$$

The unitary Ramanujan sums $c_q^*(n)$ were defined by Cohen [30] as follows:

$$c_q^*(n) = \sum_{\substack{1 \leq j \leq q \\ (j,q)_* = 1}} e(jn/q) \quad (q \in \mathbb{N}, n \in \mathbb{Z}).$$

The identities

$$c_q^*(n) = \sum_{d||(n,q)_*} d\mu^*(q/d), \tag{1.9.2}$$

$$\sum_{d||q} c_d^*(n) = \eta_q(n) := \begin{cases} q & \text{if } q \mid n, \\ 0 & \text{otherwise} \end{cases} \tag{1.9.3}$$

can be compared to the corresponding ones concerning the classical Ramanujan sums $c_q(n)$. Also, $c_q^*(n)$ is multiplicative in q for any fixed $n \in \mathbb{Z}$ and

$$c_{p^a}^*(n) = \begin{cases} p^a - 1 & \text{if } p^a \mid n, \\ -1 & \text{otherwise} \end{cases} \tag{1.9.4}$$

for any prime powers p^a $(a \geq 1)$. Note that

$$c_q^*(q) = \phi^*(q), \; c_q^*(1) = \mu^*(q) \quad (q \in \mathbb{N}).$$

If q is squarefree, then all its divisors are unitary divisors, and

$$c_q^*(n) = c_q(n).$$

There are many other unitary analogs of functions and identities known in the classical case, For example, it can be shown that

$$\sum_{n=1}^{\infty} \frac{c_q^*(n)}{n} = -\Lambda^*(q) \quad (q > 1), \tag{1.9.5}$$

where Λ^* is the unitary analog of the von Mangoldt function Λ, and is given by

$$\Lambda^*(n) = \begin{cases} a\log p & \text{if } n = p^a \text{ is a prime power } (a \geq 1), \\ 0 & \text{otherwise.} \end{cases} \tag{1.9.6}$$

Identity (1.9.5), obtained by Subbarao [105], is the analog of (1.1.4).

We will need the following results, which are the unitary analogs of (1.3.9) and (1.3.10) concerning classical Ramanujan sums.

Proposition 1.9.1 ([118, Proposition 1]). *For any* $q, n \in \mathbb{N}$,

$$\sum_{d||q} |c_d^*(n)| = 2^{\omega(q/(n,q)_*)}(n, q)_*, \tag{1.9.7}$$

$$\sum_{d||q} |c_d^*(n)| \leq 2^{\omega(q)} n. \tag{1.9.8}$$

Some other properties of Ramanujan sums and unitary Ramanujan sums differ notably. For example, the unitary sums $c_q^*(n)$ do not enjoy the orthogonality property (1.3.6) of the classical Ramanujan sums. To see this, let p be a prime, let $q_1 = p$, $q_2 = p^2$ and $n = p^2$. Then, according to (1.9.4),

$$\sum_{k=1}^{p^2} c_p^*(k)c_{p^2}^*(k) = p^2(p-1) \neq 0.$$

For further properties and generalizations of unitary Ramanujan sums, see, e.g., [30, 49, 75, 106].

We also remark that the unitary cyclotomic polynomials $\Phi_n^*(x)$ are defined by

$$\Phi_n^*(x) = \prod_{\substack{j=1 \\ (j,n)_*=1}}^{n} (x - e(j/n)). \tag{1.9.9}$$

Some of their properties, which are similar to those of the classical cyclotomic polynomials $\Phi_n(x)$, are given by Sivaramakrishnan [101, Section X.4]. Namely, $\Phi_n^*(x)$ is monic polynomial, has integer coefficients and is of degree $\phi^*(n)$. Furthermore, for any $n \in \mathbb{N}$ one has

$$x^n - 1 = \prod_{d||n} \Phi_d^*(x) \tag{1.9.10}$$

and

$$\Phi_n^*(x) = \prod_{d||n} \left(x^d - 1\right)^{\mu^*(n/d)}. \tag{1.9.11}$$

We point out the identity

$$\Phi_n^*(x) = \exp\left(-\sum_{k=1}^{\infty} \frac{c_n^*(k)}{k} x^k\right),$$

valid for any $n > 1$ and $x \in \mathbb{C}$, $|x| < 1$, which is the unitary analog of (1.1.13).

See the papers [80] and [50] for a detailed discussion on the unitary cyclotomic polynomials $\Phi_n^*(x)$.

1.9.2. Modified unitary Ramanujan sums

Let $(k, n)_{**}$ denote the greatest common unitary divisor of k and n. Note that $d \parallel (k, n)_{**}$ holds true if and only if $d \parallel k$ and $d \parallel n$. Bi-unitary analogues of the Ramanujan sums may be defined as follows:

$$c_q^{**}(n) = \sum_{\substack{1 \le k \le q \\ (k,q)_{**}=1}} e(kn/q) \quad (q, n \in \mathbb{N}).$$

However, the function $q \mapsto c_q^{**}(n)$ is not multiplicative, and its properties are not parallel to the sums $c_q(n)$ and $c_q^*(n)$. The function

$$c_q^{**}(q) = \phi^{**}(q),$$

called bi-unitary Euler function was investigated by Tóth [110].

Tóth [119] introduced for $q, n \in \mathbb{N}$ the modified unitary Ramanujan sums $\widetilde{c}_q(n)$ by the formula

$$\sum_{d \| q} \widetilde{c}_d(n) = \begin{cases} q & \text{if } q \| n, \\ 0 & \text{if } q \nmid\mid n. \end{cases} \tag{1.9.12}$$

It follows that $\widetilde{c}_q(n)$ is multiplicative in q,

$$\widetilde{c}_{p^\nu}(n) = \begin{cases} p^\nu - 1 & \text{if } p^\nu \| n, \\ -1 & \text{if } p^\nu \nmid\mid n \end{cases} \tag{1.9.13}$$

for any prime powers p^ν $(\nu \geq 1)$ and

$$\widetilde{c}_q(n) = \sum_{d \| (n,q)_{**}} d \mu^*(q/d) \quad (q, n \in \mathbb{N}). $$

The following results are analogs of (1.3.9), (1.3.10) and (1.9.7), (1.9.8), concerning the classical and unitary Ramanujan sums, respectively.

Proposition 1.9.2 ([119, Proposition 1]). *For any $q, n \in \mathbb{N}$,*

$$\sum_{d \| q} |\widetilde{c}_d(n)| = 2^{\omega(q/(n,q)_{**})} (n, q)_{**}, \tag{1.9.14}$$

$$\sum_{d \| q} |\widetilde{c}_d(n)| \leq 2^{\omega(q)} n. \tag{1.9.15}$$

Basic properties (including those mentioned above) of the classical Ramanujan sums $c_q(n)$, their unitary analogues $c_q^*(n)$ and the modified sums $\widetilde{c}_q(n)$ can be compared by Table 1.3.

1.9.3. Ramanujan sums defined by regular systems of divisors

A common generalization of the Ramanujan sums $c_q(n)$ and $c_q^*(n)$ can be defined as follows. Let $A(n)$ be a subset of the set of positive divisors of n for every $n \in \mathbb{N}$. The A-convolution of the functions $f, g : \mathbb{N} \to \mathbb{C}$ is defined by

$$(f *_A g)(n) = \sum_{d \in A(n)} f(d)g(n/d) \quad (n \in \mathbb{N}).$$

The system $A = (A(n))_{n \in \mathbb{N}}$ of divisors is called regular (cf. [83]), if the following conditions hold true:

Table 1.3. Properties of $c_q(n)$, $c_q^*(n)$ and $\tilde{c}_q(n)$.

$c_q(n)$	$c_q^*(n)$	$\tilde{c}_q(n)$
$c_q(n) = \sum_{d\mid(n,q)} d\mu(q/d)$	$c_q^*(n) = \sum_{d\mid(n,q)_*} d\mu^*(q/d)$	$\tilde{c}_q(n) = \sum_{d\|(n,q)_{**}} d\mu^*(q/d)$
$c_q(n) = \dfrac{\phi(q)\mu(q/(n,q))}{\phi(q/(n,q))}$	$c_q^*(n) = \dfrac{\phi^*(q)\mu^*(q/(n,q)_*)}{\phi^*(q/(n,q)_*)}$	$\tilde{c}_q(n) = \dfrac{\phi^*(q)\mu^*(q/(n,q)_{**})}{\phi^*(q/(n,q)_{**})}$
$c_{p^\nu}(n) = \begin{cases} p^\nu - p^{\nu-1} & \text{if } p^\nu \mid n \\ -p^{\nu-1} & \text{if } p^{\nu-1} \| n \\ 0 & \text{if } p^{\nu-1} \nmid n \end{cases}$	$c_{p^\nu}^*(n) = \begin{cases} p^\nu - 1 & \text{if } p^\nu \mid n \\ -1 & \text{if } p^\nu \nmid n \end{cases}$	$\tilde{c}_{p^\nu}(n) = \begin{cases} p^\nu - 1 & \text{if } p^\nu \| n \\ -1 & \text{if } p^\nu \nparallel n \end{cases}$
$\sum_{d\mid q} c_d(n) = \begin{cases} q & \text{if } q \mid n \\ 0 & \text{if } q \nmid n \end{cases}$	$\sum_{d\|q} c_d^*(n) = \begin{cases} q & \text{if } q \mid n \\ 0 & \text{if } q \nmid n \end{cases}$	$\sum_{d\|q} \tilde{c}_d(n) = \begin{cases} q & \text{if } q \| n \\ 0 & \text{if } q \nparallel n \end{cases}$
$\sum_{d\mid q} \lvert c_d(n)\rvert = 2^{\omega(q/(n,q))}(n,q)$	$\sum_{d\|q} \lvert c_d^*(n)\rvert = 2^{\omega(q/(n,q)_*)}(n,q)_*$	$\sum_{d\|q} \lvert \tilde{c}_d(n)\rvert = 2^{\omega(q/(n,q)_{**})}(n,q)_{**}$

(a) $(\mathcal{A}, +, *_A)$ is a commutative ring with unity,

(b) the A-convolution of multiplicative functions is multiplicative,

(c) the constant 1 function has an inverse μ_A (generalized Möbius function) with respect to $*_A$ and $\mu_A(p^a) \in \{1, 0\}$ for every prime power p^a ($a \geq 1$).

It can be shown that the system $A = (A(n))_{n \in \mathbb{N}}$ is regular if and only if

(i)

$$A(mn) = \{de : d \in A(m), e \in A(n)\}$$

for every $m, n \in \mathbb{N}$ with $(m, n) = 1$,

(ii) for every prime power p^a ($a \geq 1$) there exists a divisor $t = t_A(p^a)$ of a, called the type of p^a with respect to A, such that

$$A(p^{it}) = \{1, p^t, p^{2t}, \ldots, p^{it}\}$$

for every $i \in \{0, 1, \ldots, a/t\}$.

Examples of regular systems of divisors are $A = D$, where $D(n)$ is the set of all positive divisors of n and $A = U$, where $U(n)$ is the set of unitary divisors of n. For every prime power p^a ($a \geq 1$), one has $t_D(p^a) = 1$ and $t_U(p^a) = a$. Here $*_D$ and $*_U$ are the Dirichlet convolution and the unitary convolution, respectively. For properties of regular convolutions and related arithmetical functions, we refer to [73, 75, 83].

The following generalization of the Ramanujan sum $c_q(n)$ was introduced by McCarthy [73]. For a regular system $A = (A(q))_{q \in \mathbb{N}}$ of divisors and $n \in \mathbb{N}$ let

$$c_{A,q}(n) = \sum_{\substack{k=1 \\ (k,q)_A=1}}^{q} e(kn/q),$$

where

$$(k, q)_A = \max\{d \in \mathbb{N} : d \mid k, d \in A(q)\}.$$

Here $c_{A,q}(n)$ preserves the basic properties of $c_q(n)$. For example, for every regular system A and every $q, n \in \mathbb{N}$ one has

$$c_{A,q}(n) = \sum_{d \mid n,\, d \in A(q)} d\mu_A(q/d),$$

therefore, $c_{A,q}(n)$ is integer-valued, multiplicative in q, and

$$c_{A,q}(1) = \mu_A(q)$$

is the generalized Möbius function. The function

$$\phi_A(n) = c_{A,q}(q)$$

is the generalized Euler function.

Observe that $c_{D,q}(n) = c_q(n)$ is the classical Ramanujan sum, and $c_{U,q}(n) = c_q^*(n)$ is its unitary analog.

Let $A = (A(n))_{n \in \mathbb{N}}$ be a system of regular divisors and let $r \in \mathbb{N}$. A function f is called A-even (mod r) if $f(n) = f((n,r)_A)$ for every $n \in \mathbb{N}$. Let $\mathcal{E}_{A,r}$ denote the set of functions f which are A-even (mod r). For example, $c_{A,r}(n)$ is A-even (mod r). Let

$$\mathcal{E}_A = \bigcup_{r \in \mathbb{N}} \mathcal{E}_{A,r}.$$

The following results show that it is not possible to develop a Fourier theory concerning the generalized Ramanujan sums $c_{A_q}(n)$, analogous to usual sums $c_q(n)$. Let

$$M(f) = \lim_{x \to \infty} \frac{1}{x} \sum_{n \leq x} f(n)$$

denote the mean value of the arithmetic function f, if the limit exists.

Proposition 1.9.3 ([108, Proposition 3]). *For every regular system A,*

$$M\left(c_{A,q_1}(\cdot)c_{A,q_2}(\cdot)\right) = \begin{cases} \phi_A(q) & \text{if } q_1 = q_2 = q, \\ 0 & \text{if } q_1 q_2 > 1, \ (q_1, q_2) > 1. \end{cases}$$

However, if $A \neq D$, then there exist q_1 and q_2 such that $q_1 \neq q_2$ and

$$M\left(c_{A,q_1}(\cdot)c_{A,q_2}(\cdot)\right) \neq 0.$$

Proposition 1.9.4 ([108, Proposition 4]). *$\mathcal{E}_A$ is a vector space if and only if $A = D$.*

1.9.4. Ramanujan sums defined by regular integers (mod n)

An integer k is called regular (mod n), if there exists an integer x such that

$$k^2 x \equiv k \pmod{n},$$

i.e., the residue class of k is a regular element (in the sense of J. von Neumann) of the ring $\mathbb{Z}_n$ of residue classes (mod n). In general, an element k of a ring R is said to be (von Neumann) regular if there exists an $x \in R$

such that $k = kxk$. If every $k \in R$ has this property, then R is called a von Neumann regular ring.

Let $n > 1$ be an integer with prime factorization $n = p_1^{\nu_1} \cdots p_r^{\nu_r}$. It can be shown that $k \geq 1$ is regular $(\mathrm{mod}\ n)$ if and only if for every $i \in \{1, \ldots, r\}$ either $p_i \nmid k$ or $p_i^{\nu_i} \mid k$. These integers occur in the literature also in another context. It is said that an integer k possesses a weak order $(\mathrm{mod}\ n)$ if there exists an integer $m \geq 1$ such that

$$k^{m+1} \equiv k \ (\mathrm{mod}\ n).$$

Then the weak order of k is the smallest m with this property. It turns out that k is regular $(\mathrm{mod}\ n)$ if and only if k possesses a weak order $(\mathrm{mod}\ n)$. (see [109]).

Let

$$\mathrm{Reg}_n = \{k : 1 \leq k \leq n,\ k \text{ is regular } (\mathrm{mod}\ n)\}$$

and let $\varrho(n) = \# \mathrm{Reg}_n$ denote the number of regular integers $k\ (\mathrm{mod}\ n)$ such that $1 \leq k \leq n$. This function is multiplicative and

$$\varrho(p^\nu) = \phi(p^\nu) + 1 = p^\nu - p^{\nu-1} + 1$$

for every prime power p^ν ($\nu \geq 1$), where ϕ is the Euler function. The average order of the function $\varrho(n)$ was considered by Joshi [51]. One has

$$\sum_{n \leq x} \varrho(n) = \frac{1}{2} A x^2 + R(x),$$

where

$$A = \prod_p \left(1 - \frac{1}{p^2(p+1)}\right) = \zeta(2) \prod_p \left(1 - \frac{1}{p^2} - \frac{1}{p^3} + \frac{1}{p^4}\right) \doteq 0.8815$$

is the so called quadratic class-number constant, and

$$R(x) = O(x \log^3 x),$$

given in [51] using elementary arguments. This was improved into

$$R(x) = O(x \log x)$$

by Herzog and Smith [41], using analytic methods. Also,

$$R(x) = \Omega_\pm(x\sqrt{\log \log x})$$

(see [41]).

Haukkanen and Tóth [39] defined the analogs of Ramanujan's sums with respect to regular integers (mod q) as

$$\bar{c}_q(n) = \sum_{a \in \mathrm{Reg}_q} e(an/q) \qquad (q \in \mathbb{N}, n \in \mathbb{Z}).$$

Observe that for $n = q$, $\bar{c}_q(q) = \varrho(q)$ defined above. Also, $\bar{c}_q(1) =: \bar{\mu}(q)$ is the characteristic function of the squareful integers.

Theorem 1.9.5 ([39, Theorem 3.1]). *For every $q, n \in \mathbb{N}$,*

$$\bar{c}_q(n) = \sum_{d \| q} c_d(n).$$

Corollary 1.9.6 ([39]). (i) *For any fixed $n \in \mathbb{N}$, the function $q \mapsto c_q(n)$ is multiplicative and for any prime power p^a $(a \geq 1)$,*

$$\bar{c}_{p^a}(n) = \begin{cases} 1 + p^a - p^{a-1} & \text{if } p^a \mid n, \\ 1 - p^{a-1} & \text{if } p^{a-1} \mid n, p^a \nmid n, \\ 1 & \text{if } p^{a-1} \nmid n. \end{cases}$$

(ii) *If $q, m, n \in \mathbb{N}$ such that $(m, n) = 1$, then*

$$\bar{c}_q(m)\bar{c}_q(n) = \bar{\mu}(q)\bar{c}_q(mn). \tag{1.9.16}$$

(iii) *For fixed $q > 1$, the function $n \mapsto \bar{c}_q(n)$ is multiplicative if and only if q is squareful.*

Here identity (1.9.16) is the analog of (1.3.4) concerning the sums $c_q(n)$.

Theorem 1.9.7 ([39, Theorem 7.2]). *For every $q \in \mathbb{N}$,*

$$\sum_{n \leq x} \frac{\bar{c}_q(n)}{n} = \log x + \gamma - (\Lambda \times 1)(q) + O(x^{-1}), \tag{1.9.17}$$

where γ is Euler's constant, Λ is the von Mangoldt function and $\times$ denotes the unitary convolution.

It follows from (1.9.17) that

$$\sum_{n \leq x} \frac{\overline{c}_q(n)}{n}$$

tends to infinity as $x \to \infty$. This result may be compared to identities (1.1.4) and (1.9.5), concerning the classical and unitary Ramanujan sums, respectively.

Corollary 1.9.8 ([39, Example 3.2]). *Let $\overline{N}_r(n, k)$ denote the number of solutions of the congruence*

$$x_1 + \cdots + x_k \equiv n \pmod{r}$$

such that $x_1, \ldots, x_k$ are all regular (mod r). Then

$$\overline{N}_r(n, k) = \frac{1}{r} \sum_{d \mid r} \overline{c}_r(r/d)^k c_d(n). \tag{1.9.18}$$

Identity (1.9.18) is the analog of (1.6.14). Further properties and applications can be found in paper [39].

1.10. Ramanujan Expansions of Arithmetic Functions of Several Variables

1.10.1. Expansions of functions with respect to classical and unitary Ramanujan sums

We prove the following results.

Theorem 1.10.1 ([118, Theorem 2]). *Let $f : \mathbb{N}^k \to \mathbb{C}$ be an arithmetic function ($k \in \mathbb{N}$). Assume that*

$$\sum_{n_1, \ldots, n_k=1}^{\infty} 2^{\omega(n_1)+\cdots+\omega(n_k)} \frac{|(\mu_k * f)(n_1, \ldots, n_k)|}{n_1 \cdots n_k} < \infty. \tag{1.10.1}$$

Then, for every $n_1, \ldots, n_k \in \mathbb{N}$,

$$f(n_1, \ldots, n_k) = \sum_{q_1, \ldots, q_k=1}^{\infty} a_{q_1, \ldots, q_k} c_{q_1}(n_1) \cdots c_{q_k}(n_k), \tag{1.10.2}$$

and

$$f(n_1, \ldots, n_k) = \sum_{q_1, \ldots, q_k=1}^{\infty} a^*_{q_1, \ldots, q_k} c^*_{q_1}(n_1) \cdots c^*_{q_k}(n_k), \tag{1.10.3}$$

where

$$a_{q_1,\ldots,q_k} = \sum_{m_1,\ldots,m_k=1}^{\infty} \frac{(\mu_k * f)(m_1 q_1, \ldots, m_k q_k)}{m_1 q_1 \cdots m_k q_k},$$

$$a_{q_1,\ldots,q_k}^* = \sum_{\substack{m_1,\ldots,m_k=1 \\ (m_1,q_1)=1,\ldots,(m_k,q_k)=1}}^{\infty} \frac{(\mu_k * f)(m_1 q_1, \ldots, m_k q_k)}{m_1 q_1 \cdots m_k q_k}, \qquad (1.10.4)$$

the series (1.10.2) *and* (1.10.3) *being absolutely convergent.*

Proof. We consider the case of the unitary Ramanujan sums $c_q^*(n)$. For any $n_1, \ldots, n_k \in \mathbb{N}$ we have, by using property (1.9.3),

$$f(n_1, \ldots, n_k) = \sum_{d_1 | n_1, \ldots, d_k | n_k} (\mu_k * f)(d_1, \ldots, d_k)$$

$$= \sum_{d_1,\ldots,d_k=1}^{\infty} \frac{(\mu_k * f)(d_1, \ldots, d_k)}{d_1 \cdots d_k} \sum_{q_1 || d_1} c_{q_1}^*(n_1) \cdots \sum_{q_k || d_k} c_{q_k}^*(n_k)$$

$$= \sum_{q_1,\ldots,q_k=1}^{\infty} c_{q_1}^*(n_1) \cdots c_{q_k}^*(n_k) \sum_{\substack{d_1,\ldots,d_k=1 \\ q_1 || d_1,\ldots,q_k || d_k}}^{\infty} \frac{(\mu_k * f)(d_1, \ldots, d_k)}{d_1 \cdots d_k},$$

giving expansion (1.10.3) with the coefficients (1.10.4), by denoting

$$d_1 = m_1 q_1, \ldots, d_k = m_k q_k.$$

The rearranging of the terms is justified by the absolute convergence of the multiple series, shown as follows:

$$\sum_{q_1,\ldots,q_k=1}^{\infty} |a_{q_1,\ldots,q_k}^*| |c_{q_1}^*(n_1)| \cdots |c_{q_k}^*(n_k)|$$

$$\leq \sum_{\substack{q_1,\ldots,q_k=1 \\ m_1,\ldots,m_k=1 \\ (m_1,q_1)=1,\ldots,(m_k,q_k)=1}}^{\infty} \frac{|(\mu_k * f)(m_1 q_1, \ldots, m_k q_k)|}{m_1 q_1 \cdots m_k q_k} |c_{q_1}^*(n_1)| \cdots |c_{q_k}^*(n_k)|$$

$$= \sum_{t_1,\ldots,t_k=1}^{\infty} \frac{|(\mu_k * f)(t_1, \ldots, t_k)|}{t_1 \cdots t_k} \sum_{\substack{m_1 q_1 = t_1 \\ (m_1,q_1)=1}} |c_{q_1}^*(n_1)| \cdots \sum_{\substack{m_k q_k = t_k \\ (m_k,q_k)=1}} |c_{q_k}^*(n_k)|$$

$$\leq n_1 \cdots n_k \sum_{t_1,\ldots,t_k=1}^{\infty} 2^{\omega(t_1)+\cdots+\omega(t_k)} \frac{|(\mu_k * f)(t_1, \ldots, t_k)|}{t_1 \cdots t_k} < \infty,$$

by using inequality (1.9.8) and condition (1.10.1).

For the Ramanujan sums $c_q(n)$ the proof is along the same lines, by using inequality (1.3.10). In the case $k = 2$ the proof of (1.10.2) was given by Ushiroya [122]. $\qquad\square$

We remark that according to the generalized Wintner theorem, under conditions of Theorem 1.10.1, the mean value $M(f)$ of the function f, defined by

$$M(f) = \lim_{x_1,\ldots,x_r \to \infty} \frac{1}{x_1 \cdots x_r} \sum_{n_1 \leq x_1,\ldots,n_r \leq x_r} f(n_1,\ldots,n_r)$$

exists and

$$a_{1,\ldots,1} = a^*_{1,\ldots,1} = M(f).$$

For multiplicative functions f, condition (1.10.1) is equivalent to

$$\sum_{n_1,\ldots,n_k=1}^{\infty} \frac{|(\mu_k * f)(n_1,\ldots,n_k)|}{n_1 \cdots n_k} < \infty \qquad (1.10.5)$$

and to

$$\sum_{p\in\mathbb{P}} \sum_{\substack{\nu_1,\ldots,\nu_k=0 \\ \nu_1+\ldots+\nu_k\geq 1}}^{\infty} \frac{|(\mu_k * f)(p^{\nu_1},\ldots,p^{\nu_k})|}{p^{\nu_1+\cdots+\nu_k}} < \infty. \qquad (1.10.6)$$

We deduce the following result.

Corollary 1.10.2 ([118, **Corollary 1**]). *Let* $f : \mathbb{N}^k \to \mathbb{C}$ *be a multiplicative function* $(k \in \mathbb{N})$. *Assume that condition* (1.10.5) *or* (1.10.6) *holds. Then for every* $n_1,\ldots,n_k \in \mathbb{N}$ *one has the absolutely convergent expansions* (1.10.2), (1.10.3), *and the coefficients can be written as*

$$a_{q_1,\ldots,q_k} = \prod_{p\in\mathbb{P}} \sum_{\nu_1\geq\nu_p(q_1),\ldots,\nu_k\geq\nu_p(q_k)} \frac{(\mu_k * f)(p^{\nu_1},\ldots,p^{\nu_k})}{p^{\nu_1+\cdots+\nu_k}},$$

$$a^*_{q_1,\ldots,q_k} = \prod_{p\in\mathbb{P}} {\sum_{\nu_1,\ldots,\nu_k}}' \frac{(\mu_k * f)(p^{\nu_1},\ldots,p^{\nu_k})}{p^{\nu_1+\cdots+\nu_k}},$$

where $\sum'$ *means that for fixed* p *and* j, ν_j *takes all values* $\nu_j \geq 0$ *if* $\nu_p(q_j) = 0$, *and takes only the value* $\nu_j = 0$ *if* $\nu_p(q_j) \geq 1$.

In the special case $f(n_1, \ldots, n_k) = g((n_1, \ldots, n_k))$ we have the following theorem.

Theorem 1.10.3 ([118, Theorem 3]). *Let $g : \mathbb{N} \to \mathbb{C}$ be an arithmetic function and let $k \in \mathbb{N}$. Assume that*

$$\sum_{n=1}^{\infty} 2^{k\,\omega(n)} \frac{|(\mu * g)(n)|}{n^k} < \infty. \tag{1.10.7}$$

Then, for every $n_1, \ldots, n_k \in \mathbb{N}$,

$$g((n_1, \ldots, n_k)) = \sum_{q_1, \ldots, q_k = 1}^{\infty} a_{q_1, \ldots, q_k} c_{q_1}(n_1) \cdots c_{q_k}(n_k), \tag{1.10.8}$$

$$g((n_1, \ldots, n_k)) = \sum_{q_1, \ldots, q_k = 1}^{\infty} a^*_{q_1, \ldots, q_k} c^*_{q_1}(n_1) \cdots c^*_{q_k}(n_k) \tag{1.10.9}$$

are absolutely convergent, where

$$a_{q_1, \ldots, q_k} = \frac{1}{Q^k} \sum_{m=1}^{\infty} \frac{(\mu * g)(mQ)}{m^k}, \qquad a^*_{q_1, \ldots, q_k} = \frac{1}{Q^k} \sum_{\substack{m=1 \\ (m,Q)=1}}^{\infty} \frac{(\mu * g)(mQ)}{m^k},$$

with the notation $Q = [q_1, \ldots, q_k]$.

Proof. We apply Theorem 1.10.1. The identity

$$g((n_1, \ldots, n_k)) = \sum_{d|n_1, \ldots, d|n_k} (\mu * g)(d)$$

shows that now

$$(\mu_k * f)(n_1, \ldots, n_k) = \begin{cases} (\mu * g)(n) & \text{if } n_1 = \cdots = n_k = n, \\ 0 & \text{otherwise.} \end{cases}$$

In the unitary case the coefficients are

$$a^*_{q_1, \ldots, q_k} = \sum_{\substack{n=1 \\ m_1 q_1 = \cdots = m_k q_k = n \\ (m_1, q_1) = 1, \ldots, (m_k, q_k) = 1}}^{\infty} \frac{(\mu_k * f)(m_1 q_1, \ldots, m_k q_k)}{m_1 q_1 \cdots m_k q_k}$$

$$= \sum_{\substack{n=1 \\ q_1 \| n, \ldots, q_k \| n}}^{\infty} \frac{(\mu * g)(n)}{n^k},$$

and take into account that $q_1 \parallel n, \ldots, q_k \parallel n$ holds if and only if

$$[q_1, \ldots, q_k] = Q \parallel n,$$

that is, $n = mQ$ with $(m, Q) = 1$.

In the classical case the proof is analog. □

Let $\sigma_s(n) = \sum_{d|n} d^s$.

Corollary 1.10.4 ([118, Corollary 3]). *For every $n_1, \ldots, n_k \in \mathbb{N}$ the following series is absolutely convergent:*

$$\frac{\sigma_s((n_1, \ldots, n_k))}{(n_1, \ldots, n_k)^s} = \zeta(s+k) \sum_{q_1, \ldots, q_k=1}^{\infty} \frac{c_{q_1}(n_1) \cdots c_{q_k}(n_k)}{Q^{s+k}} \quad (s \in \mathbb{R}, s+k > 1).$$

$$(1.10.10)$$

If $s = 1$, then (1.10.10) reduces to identity (1.1.5) and if $s = 0$, then (1.10.10) gives (1.1.6). With respect to the unitary Ramanujan sums the counterparts of the above results are given in the following Corollary. Let J_s denote the Jordan function of order s given by $J_s(n) = n^s \prod_{p|n}(1 - 1/p^s)$.

Corollary 1.10.5 ([118, Corollary 4]). *For every $n_1, \ldots, n_k \in \mathbb{N}$ the following series are absolutely convergent:*

$$\frac{\sigma_s((n_1, \ldots, n_k))}{(n_1, \ldots, n_k)^s} = \zeta(s + k) \sum_{q_1, \ldots, q_k=1}^{\infty} \frac{J_{s+k}(Q)c_{q_1}^*(n_1) \cdots c_{q_k}^*(n_k)}{Q^{2(s+k)}}$$

with $s \in \mathbb{R}, s + k > 1$,

$$\frac{\sigma((n_1, \ldots, n_k))}{(n_1, \ldots, n_k)} = \zeta(k + 1) \sum_{q_1, \ldots, q_k=1}^{\infty} \frac{J_{k+1}(Q)c_{q_1}^*(n_1) \cdots c_{q_k}^*(n_k)}{Q^{2(k+1)}} \quad (k \geq 1),$$

$$(1.10.11)$$

$$d((n_1, \ldots, n_k)) = \zeta(k) \sum_{q_1, \ldots, q_k=1}^{\infty} \frac{J_k(Q)c_{q_1}^*(n_1) \cdots c_{q_k}^*(n_k)}{Q^{2k}} \quad (k \geq 2).$$

$$(1.10.12)$$

For the Jordan function J_s and for the Euler ϕ function we have the following results.

Corollary 1.10.6 ([118, Corollary 8]). *For every* $n_1, \ldots, n_k \in \mathbb{N}$ *the following series are absolutely convergent:*

$$\frac{J_s((n_1, \ldots, n_k))}{(n_1, \ldots, n_k)^s} = \frac{1}{\zeta(s+k)} \sum_{q_1, \ldots, q_k = 1}^{\infty} \frac{\mu(Q) c_{q_1}(n_1) \cdots c_{q_k}(n_k)}{J_{s+k}(Q)} \qquad (1.10.13)$$

with $s \in \mathbb{R}, s + k > 1$,

$$\frac{\phi((n_1, \ldots, n_k))}{(n_1, \ldots, n_k)} = \frac{1}{\zeta(k+1)} \sum_{q_1, \ldots, q_k = 1}^{\infty} \frac{\mu(Q) c_{q_1}(n_1) \cdots c_{q_k}(n_k)}{J_{k+1}(Q)} \qquad (k \geq 1).$$

$$(1.10.14)$$

Observe that if Q is not squarefree, then $\mu(Q) = 0$ and all terms of the sums in (1.10.13) and (1.10.14) are zero. Now if Q is squarefree, then q_j is squarefree and

$$c_{q_j}(n_j) = c_{q_j}^*(n_j)$$

for any j and any n_j. Therefore, in the expansions (1.10.13) and (1.10.14) the Ramanujan sums $c_{q_j}(n_j)$ can be replaced by their unitary analogs $c_{q_j}^*(n_j)$ for any j and any n_j (see [118, Corollary 7]).

1.10.2. Expansions of functions with respect to modified unitary Ramanujan sums

We first prove the following general result concerning the modified unitary Ramanujan sums, which is similar to Theorem 1.10.1.

Theorem 1.10.7 ([119, Theorem 2]). *Let* $f : \mathbb{N}^k \to \mathbb{C}$ *be an arithmetic function* $(k \in \mathbb{N})$. *Assume that*

$$\sum_{n_1, \ldots, n_k = 1}^{\infty} 2^{\omega(n_1) + \cdots + \omega(n_k)} \frac{|(\mu_k^* \times f)(n_1, \ldots, n_k)|}{n_1 \cdots n_k} < \infty. \qquad (1.10.15)$$

Then, for every $n_1, \ldots, n_k \in \mathbb{N}$,

$$f(n_1, \ldots, n_k) = \sum_{q_1, \ldots, q_k = 1}^{\infty} \widetilde{a}_{q_1, \ldots, q_k} \widetilde{c}_{q_1}(n_1) \cdots \widetilde{c}_{q_k}(n_k), \qquad (1.10.16)$$

where

$$\widetilde{a}_{q_1, \ldots, q_k} = \sum_{\substack{m_1, \ldots, m_k = 1 \\ (m_1, q_1) = 1, \ldots, (m_k, q_k) = 1}}^{\infty} \frac{(\mu_k^* \times f)(m_1 q_1, \ldots, m_k q_k)}{m_1 q_1 \cdots m_k q_k}, \qquad (1.10.17)$$

the series (1.10.16) *being absolutely convergent.*

Proof.   We have for any $n_1, \ldots, n_k \in \mathbb{N}$, by using property (1.9.12),

$$
f(n_1, \ldots, n_k) = \sum_{d_1 \| n_1, \ldots, d_k \| n_k} (\mu_k^* \times f)(d_1, \ldots, d_k)
$$

$$
= \sum_{d_1, \ldots, d_k = 1}^{\infty} \frac{(\mu_k^* \times f)(d_1, \ldots, d_k)}{d_1 \cdots d_k} \sum_{q_1 \| d_1} \widetilde{c}_{q_1}(n_1) \cdots \sum_{q_k \| d_k} \widetilde{c}_{q_k}(n_k)
$$

$$
= \sum_{q_1, \ldots, q_k = 1}^{\infty} \widetilde{c}_{q_1}(n_1) \cdots \widetilde{c}_{q_k}(n_k) \sum_{\substack{d_1, \ldots, d_k = 1 \\ q_1 \| d_1, \ldots, q_k \| d_k}}^{\infty} \frac{(\mu_k^* \times f)(d_1, \ldots, d_k)}{d_1 \cdots d_k},
$$

leading to expansion (1.10.16) with the coefficients (1.10.17), by denoting $d_1 = m_1 q_1, \ldots, d_k = m_k q_k$. The rearranging of the terms is justified by the absolute convergence of the multiple series, shown hereinafter:

$$
\sum_{q_1, \ldots, q_k = 1}^{\infty} |\widetilde{a}_{q_1, \ldots, q_k}| |\widetilde{c}_{q_1}(n_1)| \cdots |\widetilde{c}_{q_k}(n_k)|
$$

$$
\leq \sum_{\substack{q_1, \ldots, q_k = 1 \\ m_1, \ldots, m_k = 1 \\ (m_1, q_1) = 1, \ldots, (m_k, q_k) = 1}}^{\infty} \frac{|(\mu_k^* \times f)(m_1 q_1, \ldots, m_k q_k)|}{m_1 q_1 \cdots m_k q_k} |\widetilde{c}_{q_1}(n_1)| \cdots |\widetilde{c}_{q_k}(n_k)|
$$

$$
= \sum_{t_1, \ldots, t_k = 1}^{\infty} \frac{|(\mu_k^* \times f)(t_1, \ldots, t_k)|}{t_1 \cdots t_k} \sum_{\substack{m_1 q_1 = t_1 \\ (m_1, q_1) = 1}} |\widetilde{c}_{q_1}(n_1)| \cdots \sum_{\substack{m_k q_k = t_k \\ (m_k, q_k) = 1}} |\widetilde{c}_{q_k}(n_k)|
$$

$$
\leq n_1 \cdots n_k \sum_{t_1, \ldots, t_k = 1}^{\infty} 2^{\omega(t_1) + \cdots + \omega(t_k)} \frac{|(\mu_k^* \times f)(t_1, \ldots, t_k)|}{t_1 \cdots t_k} < \infty,
$$

by using inequality (1.9.15) and condition (1.10.15).   □

Let

$$
f(n_1, \ldots, n_k) = g((n_1, \ldots, n_k)_{*k}),
$$

where $(n_1, \ldots, n_k)_{*k}$ denotes the greatest common unitary divisor of $n_1, \ldots, n_k$. We deduce the following result, which is similar to Theorem 1.10.3.

Theorem 1.10.8 ([119, Theorem 3]). *Let $g : \mathbb{N} \to \mathbb{C}$ be an arithmetic function and let $k \in \mathbb{N}$. Assume that*

$$\sum_{n=1}^{\infty} 2^{k\,\omega(n)} \frac{|(\mu^* \times g)(n)|}{n^k} < \infty.$$

Then for every $n_1, \ldots, n_k \in \mathbb{N}$,

$$g((n_1, \ldots, n_k)_{*k}) = \sum_{q_1, \ldots, q_k = 1}^{\infty} \widetilde{a}_{q_1, \ldots, q_k} \widetilde{c}_{q_1}(n_1) \cdots \widetilde{c}_{q_k}(n_k),$$

is absolutely convergent, where

$$\widetilde{a}_{q_1, \ldots, q_k} = \frac{1}{Q^k} \sum_{\substack{m=1 \\ (m,Q)=1}}^{\infty} \frac{(\mu^* \times g)(mQ)}{m^k}, \qquad (1.10.18)$$

with the notation $Q = [q_1, \ldots, q_k]$.

Let $\sigma_s^*(n) = \sum_{d||n} d^s$, $d^*(n) = \sigma_0^*(n) = 2^{\omega(n)}$.

Corollary 1.10.9 ([119, Corollary 1]). *For every $n_1, \ldots, n_k \in \mathbb{N}$ the following series are absolutely convergent:*

$$\frac{\sigma_s^*((n_1, \ldots, n_k)_{*k})}{(n_1, \ldots, n_k)_{*k}^s}$$

$$= \zeta(s+k) \sum_{q_1, \ldots, q_k = 1}^{\infty} \frac{J_{s+k}(Q)\widetilde{c}_{q_1}(n_1) \cdots \widetilde{c}_{q_k}(n_k)}{Q^{2(s+k)}} \quad (s \in \mathbb{R}, s+k > 1),$$

$$(1.10.19)$$

$$d^*((n_1, \ldots, n_k)_{*k}) = \zeta(k) \sum_{q_1, \ldots, q_k = 1}^{\infty} \frac{J_k(Q)\widetilde{c}_{q_1}(n_1) \cdots \widetilde{c}_{q_k}(n_k)}{Q^{2k}} \quad (k \geq 2).$$

$$(1.10.20)$$

In the case $k = s = 1$ identity (1.10.19) reduces to

$$\frac{\sigma^*(n)}{n} = \zeta(2) \sum_{q=1}^{\infty} \frac{J_2(q)\widetilde{c}_q(n)}{q^4}, \qquad (1.10.21)$$

which is the analog of identity (1.1.2).

Finally, let consider the function

$$J_s^*(n) = \prod_{p^\nu \| n} (p^{s\nu} - 1),$$

representing the unitary Jordan function of order s. Here $J_s^* = \mu^* \times \mathrm{id}_s$, where $\mathrm{id}_s(n) = n^s$, and $J_1^* = \phi^*$ is the unitary Euler function, already mentioned in Section 1.9.1.

Corollary 1.10.10 ([119, Corollary 2]). *For every $n_1, \ldots, n_k \in \mathbb{N}$ the following series are absolutely convergent:*

$$\frac{J_s^*((n_1, \ldots, n_k)_{*k})}{(n_1, \ldots, n_k)_{*k}^s} = \zeta(s+k) \prod_{p \in \mathbb{P}} \left(1 - \frac{2}{p^{s+k}}\right)$$

$$\times \sum_{q_1, \ldots, q_k = 1}^{\infty} \frac{\mu^*(Q) J_{s+k}(Q) \widetilde{c}_{q_1}(n_1) \cdots \widetilde{c}_{q_k}(n_k)}{Q^{2(s+k)} \prod_{p|Q}(1 - 2/p^{s+k})}$$

$$(s \in \mathbb{R}, s + k > 1),$$

$$\frac{\phi^*((n_1, \ldots, n_k)_{*k})}{(n_1, \ldots, n_k)_{*k}} = \zeta(k+1) \prod_{p \in \mathbb{P}} \left(1 - \frac{2}{p^{k+1}}\right)$$

$$\times \sum_{q_1, \ldots, q_k = 1}^{\infty} \frac{\mu^*(Q) J_{k+1}(Q) \widetilde{c}_{q_1}(n_1) \cdots \widetilde{c}_{q_k}(n_k)}{Q^{2(k+1)} \prod_{p|Q}(1 - 2/p^{k+1})}$$

$$(k \geq 1). \tag{1.10.22}$$

Chapter 2

Cotangent Sums Related to the Estermann Zeta Function and to the Riemann Hypothesis

In this chapter, we give an overview of results of H. Maier and M. Th. Rassias on various aspects of the following cotangent sum:

$$c_0\left(\frac{r}{b}\right) := -\sum_{m=1}^{b-1} \frac{m}{b} \cot\left(\frac{\pi m r}{b}\right), \qquad (*)$$

where $r, b \in \mathbb{N}$, $b \geq 2$, $1 \leq r \leq b$ and $(r, b) = 1$.[a]

This overview consists of five sections. In the first section, we deal with questions concerning the existence of moments and equidistribution results of the cotangent sum (*). The second section is devoted to asymptotic results of these moments. A crucial role is played by continued fractions, as well as ideas from the paper of Marmi, Moussa and Yoccoz [72] on Dynamical Systems. In the third section, we describe results of the distribution of these cotangent sums (*) for arguments of special sequences and on the joint distribution for various arguments. The fourth section is devoted to the connection of the cotangent sums (*) to questions concerning the Riemann Hypothesis. It has been known for some time that these cotangent sums appear in the Nyman–Beurling criterion for the Riemann Hypothesis. In this section, several quantities believed to lead to optimal results are investigated. H. Maier and M. Th. Rassias, especially consider sums containing the Möbius function, by applying Vaughan's identity. In the fifth section, we describe work of H. Maier and M. Th. Rassias on the maximum of the cotangent sums (*) for rational numbers, as well as

---

[a] *Remark.* The symbol used for the cotangent sum (*) should not be confused with the symbol for the Ramanujan sums presented in the previous chapter.

for rational numbers with prime numerator and denominator in short intervals. The chapter concludes with six open problems related to the topics of the present work. The interested reader is also referred to the works [14–19, 32, 60–67, 69–71, 91, 95] for more details and relevant results.

2.1. The Distribution of Cotangent Sums Related to the Estermann Zeta Function

2.1.1. Introduction

Cotangent sums are associated to the zeros of the Estermann zeta function. Balasubramanian, Conrey and Heath-Brown [10] used properties of the Estermann zeta function to prove asymptotic formulas for mean-values of the product consisting of the Riemann zeta function and a Dirichlet polynomial. Period functions and families of cotangent sums appear in recent work of Bettin and Conrey (cf. [5, 13, 16]). They generalize the Dedekind sum and share with it the property of satisfying a reciprocity formula. They prove a reciprocity formula for the Vasyunin's sum [125], which appears in the Nyman–Beurling criterion for the Riemann Hypothesis.

In the present work, improvements as well as further generalizations of asymptotic formulas regarding the relevant cotangent sums are obtained. We also prove an asymptotic formula for a more general cotangent sum as well as asymptotic results and upper bounds for the moments of the cotangent sums under consideration. Furthermore, we obtain detailed information about the distribution of the values of these cotangent sums. We also establish an estimate for the rate of growth of the moments of order $2k$, as a function of k, as well as for general positive real exponents.

2.1.1.1. *The cotangent sum and its applications*

The present monograph is focused on the study of the following cotangent sum.

Definition 2.1.1.

$$c_0\left(\frac{r}{b}\right) := -\sum_{m=1}^{b-1} \frac{m}{b} \cot\left(\frac{\pi m r}{b}\right),$$

where $r, b \in \mathbb{N}$, $b \geq 2$, $1 \leq r \leq b$ and $(r, b) = 1$.

The function $c_0(r/b)$ is odd and periodic of period 1 and its value is an algebraic number.

Before presenting the main results of the work regarding this cotangent sum, we shall demonstrate its significance by exhibiting its relation to other important functions in number theory, such as the Estermann and the Riemann zeta functions, and its connections to major open problems in Mathematics, such as the Riemann Hypothesis.

Definition 2.1.2. The Estermann zeta function $E\left(s, \frac{r}{b}, \alpha\right)$ is defined by the Dirichlet series

$$E\left(s, \frac{r}{b}, \alpha\right) = \sum_{n \geq 1} \frac{\sigma_\alpha(n) \exp\left(2\pi i n r / b\right)}{n^s},$$

where $\operatorname{Re} s > \operatorname{Re} \alpha + 1$, $b \geq 1$, $(r, b) = 1$ and

$$\sigma_\alpha(n) = \sum_{d \mid n} d^\alpha.$$

It is worth mentioning that Estermann (see [35]) introduced and studied the above function in the special case when $\alpha = 0$. Much later, it was studied by Kiuchi (see [52]) for $\alpha \in (-1, 0]$.

The Estermann zeta function can be continued analytically to a meromorphic function, on the whole complex plane up to two simple poles $s = 1$ and $s = 1 + \alpha$ if $\alpha \neq 0$ or a double pole at $s = 1$ if $\alpha = 0$ (see [35, 46, 102]).

Moreover, it satisfies the functional equation:

$$E\left(s, \frac{r}{b}, \alpha\right) = \frac{1}{\pi}\left(\frac{b}{2\pi}\right)^{1+\alpha-2s} \Gamma(1-s)\Gamma(1+\alpha-s)$$

$$\times \left(\cos\left(\frac{\pi\alpha}{2}\right) E\left(1+\alpha-s, \frac{\bar{r}}{b}, \alpha\right)\right.$$

$$\left. - \cos\left(\pi s - \frac{\pi\alpha}{2}\right) E\left(1+\alpha-s, -\frac{\bar{r}}{b}, \alpha\right)\right),$$

where $\bar{r}$ is such that $\bar{r}r \equiv 1 \pmod{b}$ and $\Gamma(s)$ stands for the Gamma function.

Balasubramanian, Conrey and Heath-Brown [10] used properties of $E\left(0, \frac{r}{b}, 0\right)$ to prove an asymptotic formula for

$$I = \int_0^T \left|\zeta\left(\frac{1}{2}+it\right)\right|^2 \left|A\left(\frac{1}{2}+it\right)\right|^2 dt,$$

where $A(s)$ is a Dirichlet polynomial.

Asymptotics for functions of the form of I are useful for theorems which provide a lower bound for the portion of zeros of the Riemann zeta function $\zeta(s)$ on the critical line (see [47, 48]).

Ishibashi (see [45]) presented a nice result concerning the value of $E\left(s, \frac{r}{b}, \alpha\right)$ at $s = 0$.

Theorem 2.1.3 (Ishibashi). *Let $b \geq 2$, $1 \leq r \leq b$, $(r, b) = 1$, $\alpha \in \mathbb{N} \cup \{0\}$. Then the following hold true:*

(1) *For even α, we have*

$$E\left(0, \frac{r}{b}, \alpha\right) = \left(-\frac{i}{2}\right)^{\alpha+1} \sum_{m=1}^{b-1} \frac{m}{b} \cot^{(\alpha)}\left(\frac{\pi m r}{b}\right) + \frac{1}{4}\delta_{\alpha,0},$$

where $\delta_{\alpha,0}$ is the Kronecker delta function.

(2) *For odd α, we have*

$$E\left(0, \frac{r}{b}, \alpha\right) = \frac{B_{\alpha+1}}{2(\alpha + 1)}.$$

In the special case when $r = b = 1$, we have

$$E\left(0, 1, \alpha\right) = \frac{(-1)^{\alpha+1} B_{\alpha+1}}{2(\alpha + 1)},$$

where by B_m we denote the mth Bernoulli number, where $B_{2m+1} = 0$,

$$B_{2m} = 2\frac{(2m)!}{(2\pi)^{2m}} \sum_{\nu \geq 1} \nu^{-2m}.$$

Hence for $b \geq 2$, $1 \leq r \leq b$, $(r, b) = 1$, it follows that

$$E\left(0, \frac{r}{b}, 0\right) = \frac{1}{4} + \frac{i}{2}c_0\left(\frac{r}{b}\right),$$

where $c_0(r/b)$ is the cotangent sum (see Definition 2.1.1).

This result gives a connection between the cotangent sum $c_0(r/b)$ and the Estermann zeta function.

Period functions and families of cotangent sums appear in recent work of Bettin and Conrey [16], generalizing the Dedekind sums and sharing with it the property of satisfying a reciprocity formula. Bettin and Conrey proved the following reciprocity formula for $c_0(r/b)$:

$$c_0\left(\frac{r}{b}\right) + \frac{b}{r}c_0\left(\frac{b}{r}\right) - \frac{1}{\pi r} = \frac{i}{2}\psi_0\left(\frac{r}{b}\right),$$

where

$$\psi_0(z) = -2\frac{\log 2\pi z - \gamma}{\pi i z} - \frac{2}{\pi}\int_{\left(\frac{1}{2}\right)} \frac{\zeta(s)\zeta(1-s)}{\sin \pi s} z^{-s}\, ds,$$

and γ stands for the Euler–Mascheroni constant.

This reciprocity formula demonstrates that $c_0(r/b)$ can be interpreted as an "imperfect" quantum modular form of weight 1, in the sense of Zagier (see [14, 129]).

The cotangent sum $c_0(r/b)$ can be associated to the study of the Riemann Hypothesis, also through its relation with the so-called Vasyunin sum. The Vasyunin sum is defined as follows:

$$V\left(\frac{r}{b}\right) := \sum_{m=1}^{b-1} \left\{\frac{mr}{b}\right\} \cot\left(\frac{\pi mr}{b}\right),$$

where $\{u\} = u - \lfloor u \rfloor$, $u \in \mathbb{R}$. It can be shown (see [14, 16]) that

$$V\left(\frac{r}{b}\right) = -c_0\left(\frac{\bar{r}}{b}\right),$$

where, as mentioned previously, $\bar{r}$ is such that

$$\bar{r}r \equiv 1\,(\mathrm{mod}\ b).$$

The Vasyunin sum is itself associated to the study of the Riemann Hypothesis through the following identity (see [14, 16]):

$$\frac{1}{2\pi(rb)^{1/2}}\int_{-\infty}^{+\infty} \left|\zeta\left(\frac{1}{2} + it\right)\right|^2 \left(\frac{r}{b}\right)^{it} \frac{dt}{\frac{1}{4} + t^2}$$

$$= \frac{\log 2\pi - \gamma}{2}\left(\frac{1}{r} + \frac{1}{b}\right) + \frac{b-r}{2rb}\log\frac{r}{b} - \frac{\pi}{2rb}\left(V\left(\frac{r}{b}\right) + V\left(\frac{b}{r}\right)\right). \quad (2.1.1)$$

Note that the only non-explicit function in the right-hand side of (2.1.1) is the Vasyunin sum.

The above formula is related to the Nyman–Beurling–Baéz–Duarte–Vasyunin approach to the Riemann Hypothesis (see [8, 14]). According to this approach, the Riemann Hypothesis is true if and only if $\lim_{N \to +\infty} d_N = 0$, where

$$d_N^2 = \inf_{D_N} \frac{1}{2\pi} \int_{-\infty}^{+\infty} \left| 1 - \zeta\left(\frac{1}{2} + it\right) D_N\left(\frac{1}{2} + it\right) \right|^2 \frac{dt}{\frac{1}{4} + t^2}$$

and the infimum is taken over all Dirichlet polynomials

$$D_N(s) = \sum_{n=1}^{N} \frac{a_n}{n^s}.$$

Hence, from the above arguments it follows that from the behavior of $c_0(r/b)$, we understand the behavior of $V(r/b)$ and thus from (2.1.1) we may hope to obtain crucial information related to the Nyman–Beurling–Baéz–Duarte–Vasyunin approach to the Riemann Hypothesis.

Therefore, to sum up, one can see from all the above that the cotangent sum $c_0(r/b)$ is strongly related to important functions of Number Theory and its properties can be applied in the study of significant open problems, such as Riemann's Hypothesis.

2.1.1.2. *Main result on equidistribution*

We first state the result.

Definition 2.1.4. For $z \in \mathbb{R}$, let

$$F(z) = \text{meas}\{\alpha \in [0, 1] : g(\alpha) \leq z\},$$

where "meas" denotes the Lebesgue measure,

$$g(\alpha) = \sum_{l=1}^{+\infty} \frac{1 - 2\{l\alpha\}}{l}$$

and

$$C_0(\mathbb{R}) = \{f \in C(\mathbb{R}) : \forall \epsilon > 0, \exists \text{ a compact set } \mathcal{K} \subset \mathbb{R},$$
$$\text{such that } |f(x)| < \epsilon, \forall x \notin \mathcal{K}\}.$$

Remark. The convergence of this series has been investigated by de la Bretèche and Tenenbaum (see [23]). It depends on the partial fraction expansion of the number α.

Theorem 2.1.5.

(i) *F is a continuous function of z.*

(ii) *Let A_0, A_1 be fixed constants such that $1/2 < A_0 < A_1 < 1$. Let also*

$$H_k = \int_0^1 \left(\frac{g(x)}{\pi} \right)^{2k} dx,$$

H_k is a positive constant depending only on k, $k \in \mathbb{N}$. There is a unique positive measure μ on $\mathbb{R}$ with the following properties:

(a) *For $\alpha < \beta \in \mathbb{R}$, we have*

$$\mu([\alpha, \beta]) = (A_1 - A_0)(F(\beta) - F(\alpha)).$$

(b)

$$\int x^k d\mu = \begin{cases} (A_1 - A_0)H_{k/2} & \text{for even } k, \\ 0 & \text{otherwise.} \end{cases}$$

(c) *For all $f \in C_0(\mathbb{R})$, we have*

$$\lim_{b \to +\infty} \frac{1}{\phi(b)} \sum_{\substack{r \,:\, (r,b)=1 \\ A_0 b \le r \le A_1 b}} f\left(\frac{1}{b} c_0 \left(\frac{r}{b} \right) \right) = \int f \, d\mu,$$

where $\phi(\cdot)$ denotes the Euler phi-function.

Remark. Bruggeman (see [24, 25]) and Vardi (see [124]) have investigated the equidistribution of Dedekind sums. In contrast with the work in this paper, they consider an additional averaging over the denominator.

2.1.1.3. *Outline of the proof and further results*

In [95], Rassias proved the following asymptotic formula.

Theorem 2.1.6. *For $b \ge 2$, $b \in \mathbb{N}$, we have*

$$c_0 \left(\frac{1}{b} \right) = \frac{1}{\pi} b \log b - \frac{b}{\pi} (\log 2\pi - \gamma) + O(1).$$

The method followed in [95] is generalized in the paper [62], where some stronger results are being proved.

We initially provide a proof of an improvement of Theorem 2.1.6 as an asymptotic expansion. Namely, we prove the following theorem.

Theorem 2.1.7.   *Let $b, n \in \mathbb{N}$, $b \geq 6N$ with $N = \lfloor n/2 \rfloor + 1$. There exist absolute real constants $A_1, A_2 \geq 1$ and absolute real constants E_l, $l \in \mathbb{N}$, with $|E_l| \leq (A_1 l)^{2l}$, such that for each $n \in \mathbb{N}$ we have*

$$c_0\left(\frac{1}{b}\right) = \frac{1}{\pi}b\log b - \frac{b}{\pi}(\log 2\pi - \gamma) - \frac{1}{\pi} + \sum_{l=1}^{n} E_l b^{-l} + R_n^*(b),$$

where

$$|R_n^*(b)| \leq (A_2 n)^{4n}\, b^{-(n+1)}.$$

Additionally, we investigate the cotangent sum $c_0\left(\frac{r}{b}\right)$ for a fixed arbitrary positive integer value of r and for large integer values of b and prove the following results.

Proposition 2.1.8.   *For $r, b \in \mathbb{N}$ with $(r, b) = 1$, it holds*

$$c_0\left(\frac{r}{b}\right) = \frac{1}{r}c_0\left(\frac{1}{b}\right) - \frac{1}{r}Q\left(\frac{r}{b}\right),$$

where

$$Q\left(\frac{r}{b}\right) = \sum_{m=1}^{b-1} \cot\left(\frac{\pi m r}{b}\right)\left\lfloor \frac{rm}{b} \right\rfloor.$$

Theorem 2.1.9.   *Let $r, b_0 \in \mathbb{N}$ be fixed with $(b_0, r) = 1$. Let b denote a positive integer with $b \equiv b_0 \pmod{r}$. Then, there exists a constant $C_1 = C_1(r, b_0)$, with $C_1(1, b_0) = 0$, such that*

$$c_0\left(\frac{r}{b}\right) = \frac{1}{\pi r}b\log b - \frac{b}{\pi r}(\log 2\pi - \gamma) + C_1\, b + O(1)$$

for large integer values of b.

Remark. Theorems 2.1.7 and 2.1.9 can also follow from the results of Bettin and Conrey [16], where they proved the very interesting reciprocity formula for the cotangent sum in question.

Theorem 2.1.10.   *Let $k \in \mathbb{N}$ be fixed. Let also A_0, A_1 be fixed constants such that $1/2 < A_0 < A_1 < 1$. Then, there exist explicit constants $E_k > 0$ and $H_k > 0$, depending only on k, such that*

(a)

$$\sum_{\substack{r:(r,b)=1 \\ A_0 b \leq r \leq A_1 b}} Q\left(\frac{r}{b}\right)^{2k} = E_k \cdot (A_1^{2k+1} - A_0^{2k+1})b^{4k}\phi(b)(1+o(1)), \quad (b \to +\infty);$$

(b)

$$\sum_{\substack{r:(r,b)=1 \\ A_0 b \leq r \leq A_1 b}} Q\left(\frac{r}{b}\right)^{2k-1} = o\left(b^{4k-2}\phi(b)\right), \quad (b \to +\infty);$$

(c)

$$\sum_{\substack{r:(r,b)=1 \\ A_0 b \leq r \leq A_1 b}} c_0 \left(\frac{r}{b}\right)^{2k} = H_k \cdot (A_1 - A_0) b^{2k} \phi(b)(1 + o(1)), \quad (b \to +\infty);$$

(d)

$$\sum_{\substack{r:(r,b)=1 \\ A_0 b \leq r \leq A_1 b}} c_0 \left(\frac{r}{b}\right)^{2k-1} = o\left(b^{2k-1}\phi(b)\right), \quad (b \to +\infty).$$

Using the method of moments, we deduce detailed information about the distribution of the values of $c_0(r/b)$, where $A_0 b \leq r \leq A_1 b$ and $b \to +\infty$. Namely, we prove Theorem 2.1.5.

Finally, we study the convergence of the series $\sum_{k \geq 0} H_k x^{2k}$ and prove the following theorem.

Theorem 2.1.11. *The series*

$$\sum_{k \geq 0} H_k x^{2k}$$

converges only for $x = 0$.

2.1.2. Approximating $c_0(1/b)$ for every integer value of b

We mentioned previously the result of [95].

Proposition 2.1.12. *For every a, b, $n \in \mathbb{N}$, $b \geq 2$, with $b \nmid na$, we have*

$$x_n := \left\{\frac{na}{b}\right\} = \frac{1}{2} - \frac{1}{2b} \sum_{m=1}^{b-1} \cot\left(\frac{\pi m}{b}\right) \sin\left(2\pi mn\frac{a}{b}\right).$$

Therefore, we obtain the following proposition.

Proposition 2.1.13. *For every positive integer b, $b \geq 2$, we have*

$$c_0\left(\frac{1}{b}\right) = \frac{1}{\pi} \sum_{\substack{a \geq 1 \\ b \nmid a}} \frac{b(1 - 2x_1)}{a}.$$

Set

$$G_L(b) = \sum_{\substack{1 \le a \le L \\ b \nmid a}} \left(\frac{b}{a} \left(1 + 2 \left\lfloor \frac{a}{b} \right\rfloor \right) - 2 \right).$$

Then, we have the following lemma (see [95]).

Lemma 2.1.14. *For every b, $L \in \mathbb{N}$, with b, $L \ge 2$, we have*

$$G_L(b) = -\log \frac{L}{b} + b(\log L + \gamma) - 2L + S(L;b) + O\left(\frac{b}{L}\right),$$

where

$$S(L;b) = 2b \sum_{1 \le a \le L} \frac{1}{a} \left\lfloor \frac{a}{b} \right\rfloor.$$

The key tool for obtaining an asymptotic expansion for $S(L;b)$ is the generalized Euler summation formula. The following definition is needed.

Definition 2.1.15. The sequence B_j of Bernoulli numbers is defined by $B_{2n+1} = 0$,

$$B_{2n} = 2 \frac{(2n)!}{(2\pi)^{2n}} \sum_{\nu \ge 1} \nu^{-2n}.$$

If f is a function that is differentiable at least $(2N+1)$ times in $[0, Z]$, let

$$r_N(f, Z) = \frac{1}{(2N+1)!} \int_0^Z (u - \lfloor u \rfloor + B)^{2N+1} f^{(2N+1)}(u)du,$$

where the following notation is used:

$$(u - \lfloor u \rfloor + B)^{2N+1} = ((u - \lfloor u \rfloor) + B)^{2N+1}$$

$$:= \sum_{j=0}^{2N+1} \binom{2N+1}{j} (u - \lfloor u \rfloor)^j B_{2N+1-j}.$$

Additionally, let

$$F_i(k, b) = ((k+1)b - 1)^{-i} - (kb - 1)^{-i}.$$

Theorem 2.1.16 (Generalized Euler Summation Formula (cf. [34])). *Let f be $(2N+1)$ times differentiable in the interval $[0, Z]$. Then*

$$\sum_{\nu=0}^{Z} f(\nu) = \frac{f(0) + f(Z)}{2} + \int_0^Z f(u)du$$

$$+ \sum_{j=1}^{N} \frac{B_{2j}}{(2j)!} \left(f^{(2j-1)}(Z) - f^{(2j-1)}(0) \right) + r_N(f, Z).$$

Application of Theorem 2.1.16 yields the following results, whose proofs are given in [62].

Lemma 2.1.17. *For $N \in \mathbb{N}$, we have*

$$S(L; b) = 2b \sum_{k \leq L/b} k \left(\log \frac{(k+1)b - 1}{kb - 1} + \frac{1}{2} F_1(k, b) \right)$$

$$+ 2b \sum_{j=1}^{N} \frac{B_{2j}}{2j} \sum_{k \leq L/b} k F_{2j}(k, b) + 2b r_N \left(f, \frac{L}{b} \right),$$

where the function f satisfies:

$$f(u) = \begin{cases} 1/u & \text{if } u \geq 1, \\ 0 & \text{if } u = 0, \end{cases}$$

and $f \in C^\infty([0, \infty))$ with $f^{(j)}(0) = 0$ for $j \leq 2N + 1$.

Lemma 2.1.18. *Let*

$$r_N(b) = \sum_{l=0}^{2N+1} \binom{2N+1}{l} B_{2N+1-l} \sum_{k \leq L/b} k \, I(b, k, l),$$

where

$$I(b, k, l) = \int_{kb-1}^{(k+1)b-1} (u - \lfloor u \rfloor)^l u^{-(2N+2)} du.$$

Then, there exist absolute constants C_0, C_1 such that

$$r_N(b) = C_0 + C(N, b) 5^N (2N+1)! b^{-(2N+1)},$$

where

$$|C(N, b)| \leq C_1.$$

Lemma 2.1.19. *We have*

$$F_j(k,b) = (k+1)^{-j}b^{-j}\sum_{\nu\geq 0}\binom{-j}{\nu}(k+1)^{-\nu}b^{-\nu}$$

$$- k^{-j}b^{-j}\sum_{\nu\geq 0}\binom{-j}{\nu}k^{-\nu}b^{-\nu}.$$

Lemma 2.1.20. *Let $L, b, n \in \mathbb{N}$, $L \geq b \geq 6N$ with $N = \lfloor n/2 \rfloor + 1$. There exist absolute constants $A_1, A_2 \geq 1$, $F \in \mathbb{R}$ and absolute constants E_l, $l \in \mathbb{N}$, with $|E_l| \leq (A_1 l)^{2l}$, such that for each $n \in \mathbb{N}$ we have*

$$S(L;b) = 2L - b\log\frac{L}{b} + \log\frac{L}{b} + Fb + \gamma - 1$$

$$+ \sum_{l=1}^{n} E_l b^{-l} + R_n(b,L) + O_n\left(\frac{b^2}{L}\right),$$

where

$$|R_n(b,L)| \leq (A_2 n)^{4n}b^{-(n+1)} + O_n(1/L).$$

Therefore, we are now able to prove the following proposition.

Proposition 2.1.21. *Let $L, b, n \in \mathbb{N}$, $L \geq b \geq 6N$ with $N = \lfloor n/2 \rfloor + 1$. There exist absolute constants $A_1, A_2 \geq 1$, $F \in \mathbb{R}$ and absolute constants E_l, $l \in \mathbb{N}$, with $|E_l| \leq (A_1 l)^{2l}$, such that for each $n \in \mathbb{N}$ we have*

$$G_L(b) = b\log b + (F+\gamma)b - 1 + \sum_{l=1}^{n} E_l b^{-l} + R_n(b,L) + O_n\left(\frac{b^2}{L}\right),$$

where

$$|R_n(b,L)| \leq (A_2 n)^{4n}b^{-(n+1)} + O_n\left(\frac{1}{L}\right).$$

Proof.   It follows by putting together Lemmas 2.1.14 and 2.1.20.   □

However, by the definition of $G_L(b)$ it follows that

$$c_0\left(\frac{1}{b}\right) = \frac{1}{\pi}\lim_{L\to+\infty} G_L(b).$$

Thus by Proposition 2.1.21 we obtain the following theorem.

Theorem 2.1.22. *Let $b, n \in \mathbb{N}$, $b \geq 6N$ with $N = \lfloor n/2 \rfloor + 1$. There exist absolute constants $A_1, A_2 \geq 1$, $H \in \mathbb{R}$ and absolute constants E_l, $l \in \mathbb{N}$, with $|E_l| \leq (A_1 l)^{2l}$, such that for each $n \in \mathbb{N}$ we have*

$$c_0\left(\frac{1}{b}\right) = \frac{1}{\pi} b \log b + Hb - \frac{1}{\pi} + \sum_{l=1}^{n} E_l b^{-l} + R_n^*(b)$$

where

$$|R_n^*(b)| \leq (A_2 n)^{4n} b^{-(n+1)}.$$

By Vasyunin's theorem, we know that for sufficiently large b it holds

$$c_0\left(\frac{1}{b}\right) = \frac{1}{\pi} b \log b - \frac{b}{\pi} (\log 2\pi - \gamma) + O(\log b).$$

Therefore, by comparison of the coefficients of b in the above expressions for $c_0(1/b)$ we get

$$H = \frac{\gamma - \log 2\pi}{\pi}.$$

Hence, we obtain the following theorem, that is Theorem 2.1.7 stated in the Introduction.

Theorem 2.1.23. *Let $b, n \in \mathbb{N}$, $b \geq 6N$, with $N = \lfloor n/2 \rfloor + 1$. There exist absolute constants $A_1, A_2 \geq 1$ and absolute real constants E_l, $l \in \mathbb{N}$ with $|E_l| \leq (A_1 l)^{2l}$, such that for each $n \in \mathbb{N}$ we have*

$$c_0\left(\frac{1}{b}\right) = \frac{1}{\pi} b \log b - \frac{b}{\pi}(\log 2\pi - \gamma) - \frac{1}{\pi} + \sum_{l=1}^{n} E_l b^{-l} + R_n^*(b)$$

where

$$|R_n^*(b)| \leq (A_2 n)^{4n} b^{-(n+1)}.$$

2.1.3. Properties of $c_0\,(r/b)$ for fixed r and large b

We can generalize Proposition 2.1.12 in order to study the cotangent sum $c_0\left(\frac{r}{b}\right)$ for an arbitrary positive integer value of r as $b \to +\infty$.

Following a method similar to the one used to prove Proposition 2.1.12, one can prove the following proposition.

Proposition 2.1.24. *For every* $r, a, b, n \in \mathbb{N}, b \geq 2$, *with* $(r, b) = 1, b \nmid na$, *we have*

$$\sum_{m=1}^{b-1} \cot\left(\frac{\pi m r}{b}\right) \cos\left(2\pi m \frac{nra}{b}\right) = 0$$

and

$$x_n = \frac{1}{2} - \frac{1}{2b} \sum_{m=1}^{b-1} \cot\left(\frac{\pi m r}{b}\right) \sin\left(2\pi m \frac{nra}{b}\right).$$

Similarly to the case when $r = 1$, by the use of the identity

$$\sum_{a \geq 1} \frac{\sin(a\theta)}{a} = \frac{\pi - \theta}{2}, \; 0 < \theta < 2\pi,$$

when b is such that $(r, b) = 1$ and $b \nmid a$, we obtain

$$\sum_{\substack{a \geq 1 \\ b \nmid a}} \frac{b(1 - 2x_1)}{a} = \pi r c_0\left(\frac{r}{b}\right) + \pi \sum_{m=1}^{b-1} \cot\left(\frac{\pi m r}{b}\right) \left\lfloor \frac{rm}{b} \right\rfloor.$$

Equivalently, by Proposition 2.1.13 we have the following proposition.

Proposition 2.1.25. *For* $r, b \in \mathbb{N}$ *with* $(r, b) = 1$, *we have*

$$c_0\left(\frac{r}{b}\right) = \frac{1}{r} c_0\left(\frac{1}{b}\right) - \frac{1}{r} Q\left(\frac{r}{b}\right),$$

where

$$Q\left(\frac{r}{b}\right) = \sum_{m=1}^{b-1} \cot\left(\frac{\pi m r}{b}\right) \left\lfloor \frac{rm}{b} \right\rfloor.$$

By the use of the above proposition, we shall prove the following theorem.

Theorem 2.1.26. *Let* $r, b_0 \in \mathbb{N}$ *be fixed with* $(b_0, r) = 1$. *Let* b *denote a positive integer with* $b \equiv b_0 \pmod{r}$. *Then, there exists a constant* $C_1 = C_1(r, b_0)$, *with* $C_1(1, b_0) = 0$, *such that*

$$c_0\left(\frac{r}{b}\right) = \frac{1}{\pi r} b \log b - \frac{b}{\pi r}(\log 2\pi - \gamma) + C_1 b + O(1),$$

for large integer values of b.

Proof. By Proposition 2.1.25, we know that

$$c_0\left(\frac{r}{b}\right) = \frac{1}{r}\,c_0\left(\frac{1}{b}\right) - \frac{1}{r}Q\left(\frac{r}{b}\right).$$

However, by splitting the range of summation of $Q(r/b)$ into subintervals on which $\lfloor rm/b \rfloor$ assumes constant values, we have

$$Q\left(\frac{r}{b}\right) = \sum_{m=1}^{b-1} \cot\left(\frac{\pi m r}{b}\right)\left\lfloor\frac{rm}{b}\right\rfloor = \sum_{j=0}^{r-1} j \sum_{j \le \lfloor\frac{rm}{b}\rfloor < j+1} \cot\left(\frac{\pi m r}{b}\right).$$

We shall evaluate the inner sum by applying the partial fraction decomposition of the cotangent function. It is a known fact from Complex Analysis that

$$\pi\cot(\pi z) = \frac{1}{z} + \sum_{\substack{n=-\infty \\ n\neq 0}}^{+\infty}\left(\frac{1}{z-n} + \frac{1}{n}\right) = \frac{1}{z} + \frac{1}{z-1} + g_*(z),$$

where

$$g_*(z) = \frac{1}{z+1} + 2z\sum_{n\ge 1}\frac{1}{z^2-n^2}.$$

It follows that $g_*(z)$ is a continuously differentiable function for $0 \le z < 1$. We consider the sets

$$S_j = \{rm \; : \; bj \le rm < b(j+1), \; m \in \mathbb{Z}\}.$$

Then

$$S_j = \{bj + s_j, \; bj + s_j + r, \ldots, \; bj + s_j + d_j r\},$$

where s_j is a positive integer different from zero and d_j is an appropriate non-negative integer, since $(b, r) = 1$.
Let

$$b = s_j + d_j r + t_j \quad \text{with} \quad 1 \le t_j < r.$$

By the definition of S_j we have

$$s_j \equiv -bj \,(\mathrm{mod}\, r) \quad \text{and} \quad t_j \equiv b - s_j \,(\mathrm{mod}\, r) \tag{2.1.2}$$

and thus

$$t_j \equiv b(j+1) \,(\mathrm{mod}\, r). \tag{2.1.3}$$

By the definition of S_j and application of partial fraction decomposition, we obtain

$$\sum_{j \leq \lfloor \frac{rm}{b} \rfloor < j+1} \cot\left(\frac{\pi mr}{b}\right)$$

$$= \sum_{l=0}^{d_j} \cot\left(\pi \frac{s_j + lr}{b}\right) \text{ (since the cotangent function has period } \pi)$$

$$= \frac{b}{\pi} \sum_{l=0}^{d_j} \frac{1}{s_j + lr} + \frac{b}{\pi} \sum_{l=0}^{d_j} \frac{1}{s_j + lr - b} + \sum_{l=0}^{d_j} g_*\left(\frac{s_j + lr}{b}\right). \qquad (2.1.4)$$

We shall apply Euler's summation formula (cf. [34, p. 47]). Let f be a continuously differentiable function on the interval $[0, n]$. Then we have

$$\sum_{\nu=0}^{n} f(\nu) = \frac{f(0) + f(n)}{2} + \int_0^n f(x)dx + \int_0^n f'(x)P_1(x)dx,$$

where $P_1(x) = x - \lfloor x \rfloor - 1/2$ is the Bernoulli polynomial of first degree. We obtain

$$\sum_{l=0}^{d_j} \frac{1}{s_j + lr} = \int_0^{d_j} \frac{du}{s_j + ur} - r \int_0^{d_j} \frac{P_1(u)}{(s_j + ur)^2} du + \frac{1}{2s_j} + \frac{1}{2(s_j + d_j r)}$$

$$= \frac{1}{r} \log(s_j + d_j r) - \frac{1}{r} \log s_j - r \int_0^{+\infty} \frac{P_1(u)}{(s_j + ur)^2} du$$

$$+ \frac{1}{2s_j} + O\left(\frac{1}{b}\right). \qquad (2.1.5)$$

By the definition of S_j, we have

$$b(j+1) \leq bj + s_j + d_j r + r$$

and therefore

$$s_j + d_j r = b + O(1).$$

Analogously

$$t_j + d_j r = b + O(1).$$

By the substitution $l = d_j - \tilde{l}$ and Euler's summation formula, we obtain

$$\sum_{l=0}^{d_j} \frac{1}{s_j + lr - b} = -\sum_{\tilde{l}=0}^{d_j} \frac{1}{t_j + \tilde{l}r} \quad \text{(since } t_j = b - d_j r - s_j)$$

$$= \frac{1}{2(s_j + d_j r - b)} + \frac{1}{2(s_j - b)}$$

$$- \int_0^{d_j} \frac{du}{t_j + ur} + r\int_0^{+\infty} \frac{P_1(u)}{(t_j + ur)^2}du + O\left(\frac{1}{b}\right) \quad (2.1.6)$$

since

$$\frac{1}{2(t_j + d_j r)} = O\left(1/b\right),$$

because of the definition of S_j.

By the substitution $\nu = u/b$ and by the property $d_j = b/r + O(1)$ and Euler's summation formula, we obtain

$$\sum_{l=0}^{d_j} g_*\left(\frac{s_j + lr}{b}\right) = b\int_0^{1/r} g_*\left(vr\right)dv + O(1), \quad (2.1.7)$$

because $s_j + d_j r = b + O(1)$. Therefore by (2.1.4)–(2.1.7) we obtain

$$c_0\left(\frac{r}{b}\right) = \frac{1}{r}\,c_0\left(\frac{1}{b}\right) - \frac{1}{r}Q\left(\frac{r}{b}\right)$$

$$= \frac{1}{r}\,c_0\left(\frac{1}{b}\right) - \frac{1}{r}\sum_{j=0}^{r-1} j \sum_{j \le \lfloor \frac{rm}{b}\rfloor < j+1} \cot\left(\frac{\pi m r}{b}\right)$$

$$= \frac{1}{r}\,c_0\left(\frac{1}{b}\right)$$

$$- \frac{1}{r}\sum_{j=0}^{r-1} j \left(\frac{b}{\pi}\sum_{l=0}^{d_j} \frac{1}{s_j + lr} + \frac{b}{\pi}\sum_{l=0}^{d_j} \frac{1}{s_j + lr - b} + \sum_{l=0}^{d_j} g_*\left(\frac{s_j + lr}{b}\right)\right)$$

$$= \frac{1}{r}\,c_0\left(\frac{1}{b}\right) - \frac{b}{\pi r}\sum_{j=0}^{r-1} j\left(\frac{1}{r}\log(s_j + d_j r) - \frac{1}{r}\log s_j\right.$$

$$\left. - r\int_0^{+\infty} \frac{P_1(u)}{(s_j + ur)^2}du + \frac{1}{2s_j} + O\left(\frac{1}{b}\right)\right)$$

$$- \frac{b}{\pi r} \sum_{j=0}^{r-1} j \left(-\frac{1}{2t_j} + \frac{1}{2(s_j - b)} - (\log(t_j + rd_j) - \log t_j) \frac{1}{r} \right.$$

$$+ r \int_0^{+\infty} \frac{P_1(u)}{(t_j + ur)^2} du + O\left(\frac{1}{b}\right) \right)$$

$$- \frac{b}{r} \sum_{j=0}^{r-1} j \int_0^{1/r} g_*(vr) \, dv + O(1).$$

Thus, by Theorem 2.1.6, we obtain

$$c_0 \left(\frac{r}{b} \right) = \frac{1}{\pi r} b \log b - \frac{b}{\pi r} (\log 2\pi - \gamma) + O(1)$$

$$- \frac{b}{\pi r} \sum_{j=0}^{r-1} j \left(\frac{1}{r} \log(s_j + d_j r) - \frac{1}{r} \log s_j \right.$$

$$- r \int_0^{+\infty} \frac{P_1(u)}{(s_j + ur)^2} du + \frac{1}{2s_j} + O\left(\frac{1}{b}\right) \right)$$

$$- \frac{b}{\pi r} \sum_{j=0}^{r-1} j \left(-\frac{1}{2t_j} + \frac{1}{2(s_j - b)} - \frac{1}{r} \log(t_j + rd_j) + \frac{1}{r} \log t_j \right.$$

$$+ r \int_0^{+\infty} \frac{P_1(u)}{(t_j + ur)^2} du + O\left(\frac{1}{b}\right) \right)$$

$$- \frac{b}{r} \sum_{j=0}^{r-1} j \int_0^{1/r} g_*(vr) \, dv + O(1).$$

Thus

$$\sum_{j=0}^{r-1} \frac{j}{r} \log(s_j + d_j r) = \left(\log b + O\left(\frac{1}{b}\right) \right) \sum_{j=0}^{r-1} \frac{j}{r}, \tag{2.1.8}$$

$$\sum_{j=0}^{r-1} \frac{j}{r} \log(t_j + d_j r) = \left(\log b + O\left(\frac{1}{b}\right) \right) \sum_{j=0}^{r-1} \frac{j}{r}, \tag{2.1.9}$$

$$\sum_{j=0}^{r-1} \frac{j}{s_j - b} = O\left(\frac{1}{b}\right), \tag{2.1.10}$$

$$\sum_{j=0}^{r-1} j O\left(\frac{1}{b}\right) = O\left(\frac{1}{b}\right), \tag{2.1.11}$$

and

$$\frac{b}{r} \sum_{j=0}^{r-1} j \int_0^{1/r} g_* (vr) \, dv = kb,$$

$$(2.1.12)$$

where k is a real constant depending only upon r.

By (2.1.8)–(2.1.12), we obtain

$$
\begin{aligned}
c_0\left(\frac{r}{b}\right) &= \frac{1}{\pi r} b \log b - \frac{b}{\pi r} \log 2\pi + \frac{b}{\pi r}\gamma + O(1) \\
&\quad - \frac{b}{\pi r}\left(\log b + O\left(\frac{1}{b}\right)\right) \sum_{j=0}^{r-1} \frac{j}{r} + \frac{b}{\pi r^2} \sum_{j=0}^{r-1} j \log s_j \\
&\quad + \frac{b}{\pi} \sum_{j=0}^{r-1} j \int_0^{+\infty} \frac{P_1(u)}{(s_j + ur)^2} du - \frac{b}{2\pi r} \sum_{j=0}^{r-1} \frac{j}{s_j} - \frac{b}{\pi r} \sum_{j=0}^{r-1} j O\left(\frac{1}{b}\right) \\
&\quad + \frac{b}{2\pi r} \sum_{j=0}^{r-1} \frac{j}{t_j} - \frac{b}{2\pi r} O\left(\frac{1}{b}\right) + \frac{b}{\pi r}\left(\log b + O\left(\frac{1}{b}\right)\right) \sum_{j=0}^{r-1} \frac{j}{r} \\
&\quad - \frac{b}{\pi r^2} \sum_{j=0}^{r-1} j \log t_j - \frac{b}{\pi} \sum_{j=0}^{r-1} j \int_0^{+\infty} \frac{P_1(u)}{(t_j + ur)^2} du \\
&\quad - \frac{b}{\pi r} O\left(\frac{1}{b}\right) - \frac{b}{r} \sum_{j=0}^{r-1} j \int_0^{1/r} g_*(vr) dv + O(1).
\end{aligned}
$$

Therefore,

$$c_0\left(\frac{r}{b}\right) = \frac{1}{\pi r} b \log b - \frac{b}{\pi r}(\log 2\pi - \gamma) + C_1 b + O(1),$$

where

$$
\begin{aligned}
C_1 &= \frac{1}{\pi r^2} \sum_{j=0}^{r-1} j \log \frac{s_j}{t_j} - \frac{1}{2\pi r} \sum_{j=0}^{r-1} j \left(\frac{1}{s_j} - \frac{1}{t_j}\right) \\
&\quad + \frac{1}{\pi} \sum_{j=0}^{r-1} j \int_0^{+\infty} P_1(u) \left(\frac{1}{(s_j + ur)^2} - \frac{1}{(t_j + ur)^2}\right) du \\
&\quad - \frac{1}{r} \sum_{j=0}^{r-1} j \int_0^{1/r} g_*(vr) dv,
\end{aligned}
$$

which by (2.1.2), (2.1.3) depends only on r and b_0. This completes the proof of the theorem. $\square$

2.1.4. Moments of the cotangent sum $c_0(r/b)$ for fixed large b

A crucial feature of the sum

$$\sum_{l=0}^{d_j} \cot\left(\pi \frac{s_j + lr}{b}\right)$$

is the dominating influence of the terms $\cot(\pi\frac{s_j}{b})$, which are obtained for $l = 0$, for small values of s_j. The cause of this fact is the singularity of the function $\cot x$ at $x = 0$.

A similar influence is exercised by the terms with small values of t_j, caused by the singularity of $\cot x$ at $x = \pi$. Thus, these terms should be treated separately. The other terms may be expected to cancel, since

$$\int_\epsilon^{\pi-\epsilon} \cot x \, dx = 0,$$

coming from the functional equation

$$\cot(\pi - x) = -\cot x.$$

Because of formula (2.1.2), that is

$$s_j \equiv -bj \,(\mathrm{mod}\, r)$$

and because of formula (2.1.3), that is

$$t_j \equiv b(j + 1) \,(\mathrm{mod}\, r)$$

the quality of this cancellation will depend on good equidistribution properties of the fractions $\frac{jb}{r}$ (mod 1) for j ranging over short intervals. It is a well-known fact from Diophantine approximation that these equidistributions are only good if the fraction b/r cannot be well approximated by fractions with small denominators. Lemma 2.1.27 provides a preparation for estimating the number of such values for r.

Let A_0, A_1 be constants satisfying $1/2 < A_0 < A_1 < 1$. These constants will remain fixed throughout the section.

For $m \in \mathbb{N}$, let

$$\tilde{d}(m) := \tilde{d}(m, b)$$

denote the number of divisors r of m that satisfy

$$A_0 b \leq r \leq A_1 b, \quad (r, b) = 1.$$

Lemma 2.1.27. *Let*

$$0 < \delta \leq 1, \ \mathcal{L}_0 = b\delta, \ (s, b) = 1 \ and \ |s| \leq \mathcal{L}_0/2.$$

Then, there exists a fixed constant $M > 0$ such that

$$\sum_{l \leq \mathcal{L}_0} \tilde{d}(lb + s) \leq M\delta\phi(b),$$

where ϕ stands for the Euler totient function.

The proof of the lemma uses expansion of characteristic functions into Fourier series (see [62]).

We now establish the equidistribution properties of the fractions

$$\frac{jb}{r} \pmod 1.$$

We introduce a sequence of exceptional sets $\mathcal{E}(m)$. The quality of the equidistribution of $jb/r \pmod 1$ will be good for values of r that do not belong to an exceptional set $\mathcal{E}(m)$ with a small number m.

Lemma 2.1.28. *Let $1/2 < A_0 < A_1 < 1$. Let $\theta \in \{1, -1\}$. Let m_0 be a sufficiently large positive real constant. Let*

$$m_0 \leq m \leq 10 \log \log b.$$

Then, for all values of r such that $A_0 b \leq r \leq A_1 b$, $(b, r) = 1$ which do not belong to an exceptional set $\mathcal{E}(m)$ with

$$|\mathcal{E}(m)| = O\left(\phi(b) 2^{-m}\right),$$

the following condition holds:
Let U_1, U_2, j_1, j_2 be real numbers such that

$$U_1 \geq b^{-1} 2^{5m}, \ U_2 = U_1(1 + \delta_1), \ U_2 \leq 1,$$

where

$$j_2 - j_1 \geq b 2^{-(2m+1)}, \quad 2^{-m} \leq \delta_1 \leq 2^{-m+1}.$$

Then, we have

$$\left| \left\{ j : j_1 \leq j \leq j_2, \ \left\{ \frac{\theta j b}{r} \right\} \in [U_1, U_2] \right\} \right| = (j_2 - j_1)\delta_1 U_1 (1 + O(2^{-m})).$$

The proof of the lemma is obtained by the use of methods from Diophantine approximation.

As a preparation for the study of the dominating terms $\cot\left(\pi\frac{s_j}{b}\right)$, we now investigate an inverse problem:

How are the values of j distributed, if the value of s_j is fixed?

This requires the simultaneous localization of the values for r and its multiplicative inverses r^* (mod b). This localization will be accomplished via Fourier Analysis and upper bounds for Kloosterman sums.

Lemma 2.1.29. *Let $1/2 < A_0 < A_1 < 1$ and $r \in \mathbb{N}$. Let $\alpha \in (0,1)$, $\delta > 0$ such that $\alpha + \delta < 1$. We define*

$$b^* = b^*(r,b) \in \mathbb{N} \quad by \quad bb^* \equiv 1 \,(\mathrm{mod}\ r)$$

and

$$r^* = r^*(r,b) \in \mathbb{N} \quad by \quad rr^* \equiv 1 \,(\mathrm{mod}\ b).$$

Then, we have

$$N(\alpha,\delta) := \left| \left\{ r \, : \, r \in \mathbb{N},\ (r,b)=1,\ A_0 b \leq r \leq A_1 b,\ \alpha \leq \frac{b^*}{r} \leq \alpha + \delta \right\} \right|$$
$$= \delta(A_1 - A_0)\phi(b)(1 + o(1)), \quad (b \to +\infty).$$

The proof of the lemma uses expansion of characteristic functions into Fourier series and the estimate of Kloosterman sums.

By the use of Lemma 2.1.29 we shall prove that the sum

$$\sum_{|s_j| \leq 2^{m_1}} \cot\left(\pi\frac{s_j}{b}\right)$$

is related to the sum $f(x; m_1)$, which we define and investigate in the next two lemmas.

Lemma 2.1.30. *Let*

$$f(x; m_1) = \sum_{l=1}^{2^{m_1}} \frac{B(lx)}{l},$$

where $B(x) = 1 - 2\{x\}$. Then, for $L \in \mathbb{N}$ there are numbers $a(k,L) \in \mathbb{R}$ with

$$a(k,L) = a(k,L,m_1) = O_\epsilon(|k|^{-1+\epsilon}),$$

where the implied constant is independent from m_1, such that

$$\lim_{N \to +\infty} \left\| f(x; m_1)^L - \sum_{k=-N}^{N} a(k, L)e(kx) \right\|_2 = 0.$$

If $m_2 > m_1$, then we have

$$a(k, m_1) = a(k, m_2), \quad \text{for } |k| \le 2^{m_1}.$$

The proof of the Lemma uses induction on the exponent L.

Lemma 2.1.31. *For $f(x; m_1)$ defined as in the previous lemma, we have that the limit*

$$\lim_{m_1 \to +\infty} \int_0^1 f(x; m_1)^L dx \quad \text{exists.}$$

The proof of the lemma uses Parseval's identity and the fact that $L^1[0, 1]$ is a complete metric space.

Lemma 2.1.32. *For $x \in \mathbb{R}$, let*

$$g(x) := \sum_{l=1}^{+\infty} \frac{1 - 2\{lx\}}{l}.$$

Then, for each $x \in \mathbb{Q}$, the series $g(x)$ converges.
For $x \in \mathbb{R} \setminus \mathbb{Q}$, the series $g(x)$ converges if and only if the series

$$\sum_{m \ge 1} (-1)^m \frac{\log q_{m+1}}{q_m}$$

converges, where $(q_m)_{m \ge 1}$ denotes the sequence of partial denominators of the continued fraction expansion of x.

Proof. The statement of the lemma is part of Théorème 4.4 of the paper by de la Brétèche and Tenenbaum in [23]. □

Remark. One can show that the series $g(x)$ can also be written in the form (see [23])

$$-\sum_{l=1}^{+\infty} \frac{d(l)}{\pi l} \sin(2\pi lx),$$

where it converges, and $d(l)$ stands for the divisor function.
In the following, we will prove that the series $g(x)$ converges almost everywhere.

Definition 2.1.33. Let $\alpha \in [0,1)$ be an irrational number and $\alpha = [a_0; a_1, a_2, \ldots]$ be the continued fraction expansion of α. We denote the nth convergent of α by p_n/q_n.

Lemma 2.1.34. *Let $1 < K < \sqrt{2}$. Then, there is a positive constant $c_0 = c_0(K)$ such that $q_n \geq c_0 K^n$, for every $n \in \mathbb{N}$.*

Proof. We have

$$p_n = a_n p_{n-1} + p_{n-2}, \ p_{-1} = 1, \ p_{-2} = 0 \tag{2.1.13}$$

and

$$q_n = a_n q_{n-1} + q_{n-2}, \ q_{-1} = 0, \ q_{-2} = 1. \tag{2.1.14}$$

From (2.1.14), it follows that $q_n \geq 2q_{n-2}$, for every $n \in \mathbb{N}$. By induction on $k \in \mathbb{N}$, we conclude that

$$q_{2k} \geq q_0 2^k, \tag{2.1.15}$$

for every $k \in \mathbb{N}$. From (2.1.15) the proof of the lemma follows. $\square$

Lemma 2.1.35. *Let $\mathcal{F}_n \subseteq [0,1)$, $n \in \mathbb{N}$, be Lebesgue measurable sets such that*

$$\mathcal{F}_1 \supseteq \mathcal{F}_2 \supseteq \mathcal{F}_3 \supseteq \cdots \supseteq \mathcal{F}_n \supseteq \mathcal{F}_{n+1} \supseteq \cdots$$

Assume that

$$\sum_{i=1}^{+\infty} \operatorname{meas}(\mathcal{F}_i) < +\infty.$$

Then, we have

$$\operatorname{meas}\{\alpha \in [0,1) : \alpha \in \mathcal{F}_i \text{ for infinitely many values of } i \in \mathbb{N}\} = 0.$$

Proof. This is the Borel–Cantelli lemma (cf. [56, 89]). $\square$

Definition 2.1.36. Let $q \in \mathbb{N}$, $\delta > 0$ and $\Delta(q) := \exp(-q^\delta)$. Then we define the set

$$\mathcal{E}(q, \delta) := \bigcup_{\substack{0 \leq a \leq q \\ a \in \mathbb{Z}}} \left[\frac{a}{q} - \Delta(q), \frac{a}{q} + \Delta(q)\right].$$

Definition 2.1.37. Let $L > 1$. Then we define the set

$$\mathcal{E}(L) := \left\{\alpha \in [0,1) : \frac{\log q_{m+1}}{q_m} \geq L^{-m} \text{ for infinitely many values of } m \in \mathbb{N}\right\}.$$

Lemma 2.1.38. *There is a constant $L_0 > 1$, such that* meas $\mathcal{E}(L) = 0$ *whenever $1 < L \leq L_0$.*

Proof. By Lemma 2.1.34, we have for $1 < K < \sqrt{2}$:

$$q_m \geq c_0 K^m. \tag{2.1.16}$$

Let $0 < \delta < 1$. From (18) we obtain

$$q_m^{1-\delta} \geq c_0^{1-\delta} \left(K^{1-\delta}\right)^m. \tag{2.1.17}$$

If we choose L_0 with $1 < L_0 < K^{1-\delta}$, we get for all real values of L with $1 < L \leq L_0$ the following equation:

$$q_m^{1-\delta} \geq L^m, \text{ for } m \geq m_0, \tag{2.1.18}$$

where m_0 is a sufficiently large positive integer.

From (2.1.18) we obtain

$$L^{-m} q_m \geq q_m^\delta. \tag{2.1.19}$$

Let now $\alpha \in \mathcal{E}(L)$ and $m \geq m_0$ such that

$$\frac{\log q_{m+1}}{q_m} \geq L^{-m}. \tag{2.1.20}$$

We have

$$\frac{p_{m+1}}{q_{m+1}} - \frac{p_m}{q_m} = \frac{(-1)^{m+1}}{q_m q_{m+1}}$$

(cf. [98]).

Since α lies between p_m/q_m and p_{m+1}/q_{m+1} (cf. [98]) we have by (2.1.20) the following equation:

$$\left| \alpha - \frac{p_m}{q_m} \right| \leq \frac{1}{q_m q_{m+1}} \leq \frac{1}{q_m \exp(L^{-m} q_m)}$$

and by (2.1.19) we obtain

$$\left| \alpha - \frac{p_m}{q_m} \right| \leq \exp(-q_m^\delta).$$

Thus by Definition 2.1.36, it follows that $\alpha \in \mathcal{E}(q_m, \delta)$. By Lemma 2.1.35, we therefore have

$$\text{meas} \left\{ \alpha \in [0,1) \; : \; \frac{\log q_{m+1}}{q_m} \geq L^{-m} \text{ for infinitely many values of } m \in \mathbb{N} \right\}$$

$$\leq \text{meas} \left\{ \alpha \in [0,1) \; : \; \alpha \in \mathcal{E}(q, \delta) \text{ for infinitely many values of } q \in \mathbb{N} \right\} = 0.$$

$\square$

Lemma 2.1.39. *The series*

$$g(\alpha) = \sum_{l=1}^{+\infty} \frac{1 - 2\{l\alpha\}}{l}$$

converges almost everywhere in $[0,1)$.

Proof.   By Lemma 2.1.32, the series $g(\alpha)$ converges for each $\alpha \in [0,1)$ such that $\alpha \in \mathbb{Q}$ or $\alpha \in \mathbb{R} \setminus \mathbb{Q}$ and the series

$$\sum_{m \geq 1} (-1)^m \frac{\log q_{m+1}}{q_m} \tag{2.1.21}$$

converges. The series (2.1.21) converges if there exist $m_0 \in \mathbb{N}$ and $L > 1$ such that

$$\log q_{m+1} < L^{-m} q_m \text{ for } m \geq m_0. \tag{2.1.22}$$

By Lemma 2.1.38, equation (2.1.22) holds almost everywhere in $[0,1)$. This completes the proof of the lemma. $\square$

   Remark: The convergence of the series 2.1.18 follows from the convergence of the series

$$\sum_{m \geq 1} \frac{\log q_{m+1}}{q_m},$$

which is the defining property of the Brjuno numbers. The set of these numbers is known to have measure 1.

Theorem 2.1.40. *Let*

$$D_L := \lim_{m_1 \to +\infty} \int_0^1 f(x; m_1)^L dx.$$

For $k \in \mathbb{N}$, *we have*

$$D_{2k} = \int_0^1 g(x)^{2k} dx, \quad \text{as well as} \quad D_{2k} > 0.$$

Proof.   Since the sequence $(f(x; m_1))_{m_1 \geq 1}$ forms a Cauchy sequence in the space $L^1([0, 1])$, as it was shown in the proof of Lemma 2.1.31, there exists a limit function $w(x) \in L^1([0, 1])$ such that

$$\lim_{m_1 \to +\infty} \|f(.\,; m_1) - w(.)\|_1 = 0.$$

On the other hand, we have $f(x; m_1) \to g(x)$, almost everywhere, as $m_1 \to +\infty$. A subsequence $f(x; \nu_k)$ of the sequence $(f(x; m_1))_{m_1 \geq 1}$, $\nu_k \to +\infty$, as $k \to +\infty$, converges almost everywhere to w. Therefore, $g(x) = w(x)$, almost everywhere.

Thus, there exists a function $w_L \in L^1([0, 1])$ such that $f(.\,; m_1)^L \to w_L$, in L^1 and so $w_L(x) = g(x)^L$, almost everywhere. Hence

$$\int_0^1 f(x; m_1)^L dx \to \int_0^1 w_L(x) dx = \int_0^1 g(x)^L dx.$$

Since not all Fourier coefficients of $w(x)$ are equal to zero, we obtain

$$\int_0^1 g(x)^2 dx > 0,$$

and therefore we get

$$D_{2k} = \int_0^1 g(x)^{2k} dx > 0. \qquad \square$$

In the following we will study the moments of the sums $Q(r/b)$, which are related to the sums $c_0(r/b)$ by Proposition 2.1.25:

$$c_0 \left(\frac{r}{b} \right) = \frac{1}{r} c_0 \left(\frac{1}{b} \right) - \frac{1}{r} Q \left(\frac{r}{b} \right).$$

Here the term $\frac{1}{r} c_0 \left(\frac{1}{b} \right)$ provides only a small contribution, since by Theorem 2.1.6 we have

$$c_0 \left(\frac{1}{b} \right) = O(b \log b).$$

Thus, properties of the moments

$$\sum_{\substack{r:(r,b)=1 \\ A_0 b \leq r \leq A_1 b}} c_0 \left(\frac{r}{b} \right)^L$$

can easily be extracted from properties of the moments

$$\sum_{\substack{r:(r,b)=1 \\ A_0 b \leq r \leq A_1 b}} Q\left(\frac{r}{b}\right)^L$$

by partial summation. For the treatment of the sum

$$\sum_{\substack{r:(r,b)=1 \\ A_0 b \leq r \leq A_1 b}} Q\left(\frac{r}{b}\right)^L,$$

we make use of the preparations made in Lemmas 2.1.28 and 2.1.29. From the sum

$$Q\left(\frac{r}{b}\right) = \sum_{j=0}^{r-1} j \sum_{l=0}^{d_j} \cot\left(\pi \frac{s_j + lr}{b}\right),$$

we split off the terms with $l = 0$ and small values of s_j as well as the terms with small values of t_j. The resulting sum which provides the main contribution is approximated by the sum $Q(r, b, m_1)$ (defined in (4.41) of [62]), which depends on $\alpha = b^*/r$. We shall use the localization of $\alpha = b^*/r$ established in Lemma 2.1.29. For the remaining terms of the sum $Q(r/b)$ we make use of their cancellation, using the results of Lemma 2.1.28. We obtain the following theorem (see [62]).

Theorem 2.1.41. *Let $k \in \mathbb{N}$ be fixed. Let also A_0, A_1 be fixed constants such that $1/2 < A_0 < A_1 < 1$. Then there exists a constant $E_k > 0$, depending only on k, such that*

$$\sum_{\substack{r:(r,b)=1 \\ A_0 b \leq r \leq A_1 b}} Q\left(\frac{r}{b}\right)^{2k} = E_k \cdot (A_1^{2k+1} - A_0^{2k+1}) b^{4k} \phi(b)(1 + o(1)), \quad (b \to +\infty),$$

with

$$E_k = \frac{D_{2k}}{(2k+1)\pi^{2k}}.$$

Theorem 2.1.42. *Let $k \in \mathbb{N}$ be fixed. Let also A_0, A_1 be fixed constants such that $1/2 < A_0 < A_1 < 1$. Then there exists a constant $H_k > 0$, depending only on k, such that*

$$\sum_{\substack{r:(r,b)=1 \\ A_0 b \leq r \leq A_1 b}} c_0 \left(\frac{r}{b}\right)^{2k} = H_k \cdot (A_1 - A_0) b^{2k} \phi(b)(1 + o(1)), \quad (b \to +\infty).$$

Proof. From Proposition 2.1.25 for $r, b \in \mathbb{N}$ with $(r, b) = 1$, it holds

$$c_0\left(\frac{r}{b}\right) = \frac{1}{r} c_0\left(\frac{1}{b}\right) - \frac{1}{r} Q\left(\frac{r}{b}\right).$$

Applying Theorem 2.1.6, we obtain

$$c_0\left(\frac{r}{b}\right) = -\frac{1}{r} Q\left(\frac{r}{b}\right) + O(\log b).$$

By the binomial theorem, we get

$$\sum_{\substack{r:(r,b)=1 \\ A_0 b \leq r \leq A_1 b}} c_0\left(\frac{r}{b}\right)^{2k} = \sum_{\substack{r:(r,b)=1 \\ A_0 b \leq r \leq A_1 b}} \left(\frac{Q\left(\frac{r}{b}\right)}{r}\right)^{2k}$$

$$+ O\left(\sum_{l=1}^{2k} \binom{2k}{l} \sum_{\substack{r:(r,b)=1 \\ A_0 b \leq r \leq A_1 b}} \left|Q\left(\frac{r}{b}\right)\right|^{2k-l} (\log b)^l\right).$$

$$(2.1.23)$$

By Hölder's inequality, we get

$$\sum_{\substack{r:(r,b)=1 \\ A_0 b \leq r \leq A_1 b}} \left|\frac{Q\left(\frac{r}{b}\right)}{r}\right|^{2k-l} \leq \left(\sum_{\substack{r:(r,b)=1 \\ A_0 b \leq r \leq A_1 b}} \left|\frac{Q\left(\frac{r}{b}\right)}{r}\right|^{2k}\right)^{(2k-l)/2k} \left(\sum_{\substack{r:(r,b)=1 \\ A_0 b \leq r \leq A_1 b}} 1\right)^{l/2k}.$$

Therefore (using formula (4.131) of [62]), we have

$$\sum_{\substack{r:(r,b)=1 \\ A_0 b \leq r \leq A_1 b}} \left|\frac{Q\left(\frac{r}{b}\right)}{r}\right|^{2k-l} = O\left(b^{2k-l}\phi(b)\right). \qquad (2.1.24)$$

From (2.1.23) and (2.1.24), we obtain

$$\sum_{\substack{r:(r,b)=1 \\ A_0 b \leq r \leq A_1 b}} c_0\left(\frac{r}{b}\right)^{2k} = \sum_{\substack{r:(r,b)=1 \\ A_0 b \leq r \leq A_1 b}} \left(\frac{Q\left(\frac{r}{b}\right)}{r}\right)^{2k} + O\left(b^{2k-1}\phi(b)\right). \qquad (2.1.25)$$

Using Abel's partial summation, it follows that

$$\sum_{\substack{r:(r,b)=1 \\ A_0 b \leq r \leq A_1 b}} c_0 \left(\frac{r}{b}\right)^{2k} = (A_1 b)^{-2k} \sum_{\substack{r:(r,b)=1 \\ A_0 b \leq r \leq A_1 b}} Q\left(\frac{r}{b}\right)^{2k}$$

$$+ 2k \int_{A_0 b}^{A_1 b} u^{-(2k+1)} \sum_{\substack{r:(r,b)=1 \\ A_0 b \leq r \leq u}} Q\left(\frac{r}{b}\right)^{2k} du. \quad (2.1.26)$$

By Theorem 2.2.9, we obtain

$$\sum_{\substack{r:(r,b)=1 \\ A_0 b \leq r \leq u}} Q\left(\frac{r}{b}\right)^{2k} = E_k \cdot \left(\left(\frac{u}{b}\right)^{2k+1} - A_0^{2k+1}\right) b^{4k} \phi(b)(1+o(1)). \quad (2.1.27)$$

From (2.1.26) and (2.1.27), we get

$$\sum_{\substack{r:(r,b)=1 \\ A_0 b \leq r \leq A_1 b}} c_0 \left(\frac{r}{b}\right)^{2k} = E_k \cdot (A_1 b)^{-2k} \left(A_1^{2k+1} - A_0^{2k+1}\right) b^{4k} \phi(b)(1+o(1))$$

$$+ 2k E_k \cdot \left(\int_{A_0 b}^{A_1 b} u^{-(2k+1)} \left(\left(\frac{u}{b}\right)^{2k+1} - A_0^{2k+1}\right) du\right)$$

$$\times b^{4k} \phi(b)(1+o(1)). \quad (2.1.28)$$

If we make the substitution $v = u/b$ in (2.1.28), we get

$$\sum_{\substack{r:(r,b)=1 \\ A_0 b \leq r \leq A_1 b}} c_0 \left(\frac{r}{b}\right)^{2k}$$

$$= E_k \cdot A_1^{-2k}(A_1^{2k+1} - A_0^{2k+1})b^{2k}\phi(b)(1+o(1))$$

$$+ 2k E_k \cdot \left(\int_{A_0}^{A_1} v^{-(2k+1)}(v^{2k+1} - A_0^{2k+1})dv\right) b^{2k}\phi(b)(1+o(1))$$

$$= E_k \cdot \left(A_1 - A_1^{-2k} A_0^{2k+1}\right) b^{2k}\phi(b)(1+o(1))$$

$$+ 2k E_k \cdot \left(\int_{A_0}^{A_1} \left(1 - A_0^{2k+1} v^{-(2k+1)}\right) dv\right) b^{2k}\phi(b)(1+o(1))$$

$$= (2k+1)E_k \cdot (A_1 - A_0) b^{2k}\phi(b)(1+o(1)), \quad (b \to +\infty).$$

Theorem 2.1.42, that is part (c) of Theorem 2.1.10, now follows by setting

$$H_k = (2k+1)E_k.$$

Remark. From the above theorem it follows that

$$H_k = \frac{D_{2k}}{\pi^{2k}} = \int_0^1 \left(\frac{g(x)}{\pi}\right)^{2k} dx,$$

where

$$g(x) = \sum_{l=1}^{+\infty} \frac{1 - 2\{lx\}}{l}.$$

$\square$

Theorem 2.1.43. *Let* $k \in \mathbb{N}$ *be fixed. Let also* A_0, A_1 *be fixed constants such that* $1/2 < A_0 < A_1 < 1$. *Then we have*

$$\sum_{\substack{r:(r,b)=1 \\ A_0 b \leq r \leq A_1 b}} c_0 \left(\frac{r}{b}\right)^{2k-1} = o\left(b^{2k-1}\phi(b)\right), \quad (b \to +\infty).$$

Proof. In the formulas (2.1.23), (2.1.24) and (2.1.25) from the proof of Theorem 2.1.42 we replace the exponent $2k$ by $2k - 1$ and obtain

$$\sum_{\substack{r:(r,b)=1 \\ A_0 b \leq r \leq A_1 b}} c_0 \left(\frac{r}{b}\right)^{2k-1} = \sum_{\substack{r:(r,b)=1 \\ A_0 b \leq r \leq A_1 b}} \left(\frac{Q\left(\frac{r}{b}\right)}{r}\right)^{2k-1} + O(b^{2k-2}\phi(b)).$$

$$(2.1.29)$$

Using Abel's partial summation, we get

$$\sum_{\substack{r:(r,b)=1 \\ A_0 b \leq r \leq A_1 b}} c_0 \left(\frac{r}{b}\right)^{2k-1} = (A_1 b)^{-(2k-1)} \sum_{\substack{r:(r,b)=1 \\ A_0 b \leq r \leq A_1 b}} Q\left(\frac{r}{b}\right)^{2k-1}$$

$$+ (2k - 1) \int_{A_0 b}^{A_1 b} u^{-2k} \sum_{\substack{r:(r,b)=1 \\ A_0 b \leq r \leq u}} Q\left(\frac{r}{b}\right)^{2k-1} du.$$

$$(2.1.30)$$

By Theorem 2.1.41, we obtain

$$\sum_{\substack{r:(r,b)=1 \\ A_0 b \leq r \leq u}} Q\left(\frac{r}{b}\right)^{2k-1} = o(b^{4k-2}\phi(b)), \quad (b \to +\infty). \qquad (2.1.31)$$

Thus, Theorem 2.1.43 (that is part (d) of Theorem 2.1.10) follows from the formulas (2.1.30) and (2.1.31) by substitution. $\square$

2.1.5. Probabilistic distribution

Definition 2.1.44. For $z \in \mathbb{R}$, let

$$F(z) = \text{meas}\{\alpha \in [0,1] : g(\alpha) \leq z\}$$

with

$$g(\alpha) = \sum_{l=1}^{+\infty} \frac{1 - 2\{l\alpha\}}{l}$$

and

$$C_0(\mathbb{R}) = \{f \in C(\mathbb{R}) : \forall \epsilon > 0, \exists \text{ a compact set } \mathcal{K} \subset \mathbb{R},$$
$$\text{such that } |f(x)| < \epsilon, \forall x \notin \mathcal{K}\},$$

where "meas" denotes the Lebesgue measure.

Theorem 2.1.45.

i) *F is a continuous function of z.*
ii) *Let A_0, A_1 be fixed constants, such that $1/2 < A_0 < A_1 < 1$. Let also*

$$H_k = \int_0^1 \left(\frac{g(x)}{\pi}\right)^{2k} dx.$$

There is a unique positive measure μ on $\mathbb{R}$ with the following properties:

(a) *For $\alpha < \beta \in \mathbb{R}$ we have*

$$\mu([\alpha, \beta]) = (A_1 - A_0)(F(\beta) - F(\alpha)).$$

(b)

$$\int x^k d\mu = \begin{cases} (A_1 - A_0)H_{k/2} & \text{for even } k, \\ 0 & \text{otherwise.} \end{cases}$$

(c) *For all $f \in C_0(\mathbb{R})$, we have*

$$\lim_{b \to +\infty} \frac{1}{\phi(b)} \sum_{\substack{r \, : \, (r,b)=1 \\ A_0 b \leq r \leq A_1 b}} f\left(\frac{1}{b} c_0\left(\frac{r}{b}\right)\right) = \int f \, d\mu,$$

where $\phi(\cdot)$ denotes the Euler phi-function.

Definition 2.1.46. A distribution function G is a monotonically increasing function $G\colon\mathbb{R} \to [0,1]$. The characteristic function ψ of G is defined by the following Stieltjes integral:

$$\psi(t) = \int_{-\infty}^{+\infty} e^{itu} dG(u). \quad (\text{cf. } [33, \text{ p. } 27])$$

Lemma 2.1.47. *The distribution function G is continuous if and only if the characteristic function ψ satisfies*

$$\liminf_{T\to+\infty} \frac{1}{2T} \int_{-T}^{T} |\psi(t)|^2 dt = 0.$$

Proof. See [33, Lemma 1.23, p. 48]. $\qquad\square$

We now make preparations for an application of Lemma 2.1.47 with $G = F$, and

$$\psi(t) = \Phi(t) := \int_0^1 e\left(\frac{tg(\alpha)}{2\pi}\right) d\alpha.$$

Lemma 2.1.48. *The function $h(\alpha)$ has a Fourier expansion*

$$h(\alpha) = \sum_{n>K} c(n) \sin(2\pi n\alpha),$$

with

$$|c(n)| \leq \frac{2d(n)}{\pi n},$$

where d stands for the divisor function.

Lemma 2.1.49. *We have*

$$\lim_{t\to+\infty} \Phi(t) = \lim_{t\to-\infty} \Phi(t) = 0.$$

The proof of the lemma uses a delicate partition of the interval $[0,1]$ of integration into subintervals characterized by properties of the integrand. For details see [62].

Lemma 2.1.50. *F is a continuous function of z.*

Proof. This follows from Lemmas 2.1.47 and 2.1.49. Thus, part (i) of Theorem 2.1.45 is now proved. $\qquad\square$

In the following we will prove part (ii) of Theorem 2.1.45.

Definition 2.1.51. Let $f : \mathbb{R} \to \mathbb{R}$. We set

$$\Lambda(f, b) := \frac{1}{\phi(b)} \sum_{\substack{r:(r,b)=1 \\ A_0 b \leq r \leq A_1 b}} f\left(\frac{1}{b} c_0 \left(\frac{r}{b}\right)\right).$$

We also set

$$\Lambda(f) := \lim_{b \to +\infty} \frac{1}{\phi(b)} \sum_{\substack{r:(r,b)=1 \\ A_0 b \leq r \leq A_1 b}} f\left(\frac{1}{b} c_0 \left(\frac{r}{b}\right)\right),$$

for all f for which the right-hand side exists in $\mathbb{R}$.

Lemma 2.1.52. *Let $\alpha < \beta \in \mathbb{R}$, $I = [\alpha, \beta)$. The characteristic function $\chi(.\,; I)$ is defined by*

$$\chi(u; I) = \begin{cases} 1 & \text{if } u \in I, \\ 0 & \text{otherwise.} \end{cases}$$

Then

$$\Lambda(\chi) = (A_1 - A_0)(F(\beta) - F(\alpha)) = (A_1 - A_0) \int_\alpha^\beta \chi(u; I) dF(u).$$

Like in the proof of Theorem 2.2.9 we use localization of $\alpha = \frac{b^*}{r}$ established in Lemma 2.1.29.

Lemma 2.1.53. *Let $f \in C_c(\mathbb{R})$, where $C_c(\mathbb{R})$ denotes the space of all continuous functions $f : \mathbb{R} \to \mathbb{R}$ with compact support equipped with the sup-norm. Then, we have*

$$\Lambda(f) = (A_1 - A_0) \int_{-\infty}^{+\infty} f(u) dF(u).$$

The map $\Lambda: f \to \Lambda(f)$ is a bounded linear functional on $C_c(\mathbb{R})$.

For the proof of the lemma, we approximate the function f by a linear combination of characteristic functions and apply Lemma 2.1.52.
A generalization of Definition 2.1.44 is the following definition.

Definition 2.1.54. Let X be a locally compact Hausdorff space. We set

$$C_0(X) = \{f : X \to \mathbb{R}, \, f \in C(X), \, \forall \epsilon > 0, \, \exists \text{ a compact set } \mathcal{K} \subseteq X,$$
$$\text{such that } |f(u)| < \epsilon, \, \forall u \notin \mathcal{K}\}.$$

Lemma 2.1.55. *There is a unique positive measure μ on $\mathbb{R}$ with the following properties:*

(a) *For $\alpha < \beta \in \mathbb{R}$, we have*

$$\mu([\alpha, \beta]) = (A_1 - A_0)(F(\beta) - F(\alpha)).$$

(b)

$$\int x^k d\mu = \begin{cases} (A_1 - A_0)H_{k/2} & \text{for even } k, \\ 0 & \text{otherwise.} \end{cases}$$

(c) *For all $f \in C_0(\mathbb{R})$, we have*

$$\lim_{b \to +\infty} \frac{1}{\phi(b)} \sum_{\substack{r \, : \, (r,b)=1 \\ A_0 b \le r \le A_1 b}} f\left(\frac{1}{b}c_0\left(\frac{r}{b}\right)\right) = \int f \, d\mu,$$

where $\phi(\cdot)$ denotes the Euler phi-function.

Proof. By Lemma 2.1.53 we know that Λ is a positive bounded linear functional on $C_c(\mathbb{R})$. Since $C_c(\mathbb{R})$ is dense in $C_0(\mathbb{R})$, with respect to the supremum norm, the functional Λ may be extended in a unique way to $C_0(\mathbb{R})$.

By the Riesz representation theorem, there is a unique measure μ on $\mathbb{R}$, with

$$\Lambda(f) = \int_{\mathbb{R}} f d\mu,$$

for every $f \in C_0(\mathbb{R})$. This proves (c).

Due to Lemma 2.1.52 we have

$$\mu([\alpha, \beta]) = (A_1 - A_0)(F(\beta) - F(\alpha)).$$

It follows that μ is positive. This proves (a).

Proof of (b): For every $A \in (0, +\infty)$, we set

$$g_A(u) = \begin{cases} g(u) & \text{if } |g(u)| < A, \\ 0 & \text{otherwise.} \end{cases}$$

By the definition of the Lebesgue integral, for $k \in \mathbb{N}$ we have

$$\int_0^1 g_A(u)^k du = \int_{-A}^A x^k dF(x).$$

We define $\varphi(x) := x^k$ and

$$\varphi_A(x) := \begin{cases} x^k & \text{if } |x| \leq A, \\ 0 & \text{if } |x| > A. \end{cases}$$

Since the function $\varphi_A(x)$ has compact support, we conclude from (c) that

$$\Lambda(\varphi_A) = \int \varphi_A(x) \, d\mu.$$

We choose a sequence (A_n) of real numbers with $\lim_{n \to +\infty} A_n = +\infty$. By Theorem 2.1.40 we know that

$$\int_0^1 g(u)^{2k} du$$

exists for every $k \in \mathbb{N}$. By the Cauchy–Schwarz inequality, we get

$$\int_0^1 |g(u)|^k du \leq \left(\int_0^1 g(u)^{2k} du \right)^{1/2}$$

and

$$g_{A_n}(u)^k \leq |g(u)|^k.$$

By Lebesgue's dominated convergence theorem (cf. [99, Theorem 1.34, p. 27]) we have

$$\int x^k d\mu = \int_{-\infty}^{+\infty} x^k dF(x)$$

$$= \lim_{n \to +\infty} \int_0^1 g_{A_n}(u)^k du$$

$$= \int_0^1 \left(\lim_{n \to +\infty} g_{A_n}(u)^k \right) du$$

$$= \int_0^1 g(u)^k du$$

and thus by Theorems 2.1.42 and 2.1.43, as well as Definition 2.1.51 for $f(x) = x^k$ we get

$$\lim_{b \to +\infty} \frac{1}{\phi(b)} \sum_{\substack{r \, : \, (r,b)=1 \\ A_0 b \le r \le A_1 b}} \left(\frac{1}{b} c_0 \left(\frac{r}{b} \right) \right)^k = \Lambda(f) = \int_0^1 g(u)^k du$$

$$= \begin{cases} (A_1 - A_0) H_{k/2} & \text{for even } k, \\ 0 & \text{otherwise.} \end{cases}$$

This proves (b). Therefore, the lemma is proved. $\qquad\square$

Proof of Theorem 2.1.45: The theorem now follows from Lemmas 2.1.50 and 2.1.55.

2.1.5.1. *Radius of convergence*

Theorem 2.1.56. *The series*

$$\sum_{k \ge 0} H_k x^{2k},$$

where

$$H_k = \int_0^1 \left(\frac{g(x)}{\pi} \right)^{2k} dx$$

with

$$g(x) = \sum_{l=1}^{+\infty} \frac{1 - 2\{lx\}}{l},$$

converges only for $x = 0$.

Definition 2.1.57. For $k \in \mathbb{N} \cup \{0\}$, we set

$$I := I(k) = \left[e^{-2k-1}, e^{-2k} \right] \quad \text{and} \quad l_0 := l_0(k) = e^{2k}.$$

We fix $\delta > 0$ sufficiently small and set

$$g_1(\alpha) := \sum_{l \le l_0^{1-2\delta}} \frac{B(l\alpha)}{l},$$

$$g_2(\alpha) := \sum_{l_0^{1-2\delta} < l \le l_0^{1+2\delta}} \frac{B(l\alpha)}{l},$$

$$g_3(\alpha) := \sum_{l > l_0^{1+2\delta}} \frac{B(l\alpha)}{l},$$

where

$$B(u) = 1 - 2\{u\}, \quad u \in \mathbb{R}.$$

In the sequel, we assume $k \geq k_0$ sufficiently large.

Lemma 2.1.58. *We have*

$$g(\alpha) = g_1(\alpha) + g_2(\alpha) + g_3(\alpha),$$

for every $\alpha \in \mathbb{R}$.

Proof. It is obvious by the definition of $g(\alpha)$, $g_1(\alpha)$, $g_2(\alpha)$, $g_3(\alpha)$. $\square$

Lemma 2.1.59. *For $\alpha \in I$, we have*

$$g_1(\alpha) \geq \frac{k}{2}, \quad for \ k \in \mathbb{N} \cup \{0\}.$$

Proof. For $\alpha \in I$, $l \leq l_0^{1-2\delta}$ we have $l\alpha \leq 1/4$ and therefore

$$B(l\alpha) \geq \frac{1}{2}.$$

Thus

$$g_1(\alpha) \geq \frac{1}{2} \sum_{l \leq l_0^{1-2\delta}} \frac{1}{l} \geq \frac{k}{2}.$$

$\square$

Lemma 2.1.60. *It holds*

$$|g_2(\alpha)| \leq 8\delta k, \quad for \ k \in \mathbb{N} \cup \{0\}$$

and sufficiently small $\delta > 0$.

Proof. We have

$$
\begin{aligned}
|g_2(\alpha)| &\leq \sum_{l_0^{1-2\delta} < l \leq l_0^{1+2\delta}} \frac{1}{l} \\
&\leq 2 \left(\log(l_0^{1+2\delta}) - \log(l_0^{1-2\delta}) \right) \\
&\leq 8\delta k.
\end{aligned}
$$

$\square$

Lemma 2.1.61. *For all $\alpha \in I$ that do not belong to an exceptional set $\mathcal{E}$ with measure*

$$\mathrm{meas}(\mathcal{E}) \leq e^{-2k(1+\delta)},$$

we have

$$|g_3(\alpha)| \leq \frac{1}{8}k.$$

Proof. The function g_3 has the Fourier expansion:

$$g_3(\alpha) = \sum_{l > l_0^{1+2\delta}} c(l)e(l\alpha),$$

where $c(l) = O(l^{-1+\epsilon})$ for ϵ arbitrarily small, by Lemma 2.1.48. By Parseval's identity, we have

$$\int_0^1 g_3(\alpha)^2 d\alpha = \sum_{l > l_0^{1+2\delta}} c(l)^2 = O\left(\sum_{l > l_0^{1+2\delta}} l^{-2+2\epsilon}\right) = O(l_0^{-1-3\delta/2}).$$

This completes the proof of the lemma. $\qquad\square$

Proof of Theorem 2.1.56. By Lemmas 2.1.58, 2.1.60 and 2.1.61, we have

$$|g(\alpha)| \geq |g_1(\alpha)| - |g_2(\alpha)| - |g_3(\alpha)| \geq \frac{k}{4},$$

for all $\alpha \in I$ except for those α that belong to an exceptional set

$$\mathcal{E}(I) := \mathcal{E} \cap I \subset I$$

with

$$\mathrm{meas}(\mathcal{E}(I)) \leq \frac{1}{2}|I|,$$

where $|I|$ stands for the length of I. Hence, we obtain

$$H_k = \int_0^1 \left(\frac{g(\alpha)}{\pi}\right)^{2k} d\alpha \geq \frac{1}{2}|I| \left(\frac{k}{4\pi}\right)^{2k} \geq e^{k \log k}.$$

Therefore

$$\lim_{k \to +\infty} H_k^{1/k} = +\infty$$

and thus the series

$$\sum_{k \geq 0} H_k x^{2k}$$

converges only for $x = 0$. This completes the proof of Theorem 2.1.56. $\quad\square$

Remark. In subsequent work (see [60, 61]) we settled the question of the order of magnitude for moments of the cotangent sums under consideration.

2.2. Moments of Cotangent Sums Related to the Estermann Zeta Function

In the papers [60, 61, 63, 64], Maier and Rassias continue the investigations of their papers [62, 95]. In the papers [62, 95], Maier and Rassias established the existence of the moments

$$H_k = \lim_{b \to +\infty} \phi(b)^{-1} b^{-2k} (A_1 - A_0)^{-1} \sum_{\substack{A_0 b \leq r \leq A_1 b \\ (r,b)=1}} c_0 \left(\frac{r}{b}\right)^{2k}, \quad k \in \mathbb{N},$$

where $\phi(\cdot)$ denotes the Euler phi-function. In the papers [60, 61, 63, 64], they now investigate the size of these moments as a function of k. Maier and Rassias prove the following theorem.

Theorem 2.2.1. *There exists a constant $C_0 > 0$ such that*

$$\int_0^1 |g(x)|^L dx \leq C_0^L L^L, \quad \text{for all } L \in \mathbb{N}.$$

Theorem 2.2.2. *The series*

$$\sum_{k \geq 0} \frac{H_k}{(2k)!} x^k$$

diverges for $|x| > \pi^2$, where $x \in \mathbb{C}$.

2.2.1. Continued fractions

The main ideas of the proofs of the theorems are contained in the paper [23] which belong to the field of continued fractions. We shortly describe them.

Definition 2.2.3. Let $\alpha \in [0,1] \setminus \mathbb{Q}$. Assume that

$$\alpha = [0; a_1, a_2, \ldots] = \cfrac{1}{a_1 + \cfrac{1}{a_2 + \cfrac{1}{a_3 + \cdots}}}$$

is its continued fraction expansion with integers $a_i \geq 1$ for $i = 1, 2, \ldots$ We denote the partial quotients by p_r/q_r, i.e.,

$$[0; a_1, a_2, \ldots, a_r] = \frac{p_r}{q_r}, \quad \text{with } (p_r, q_r) = 1.$$

We set $p_{-1} = 1$, $q_{-1} = 0$, $p_0 = 0$, $q_0 = 1$.

Definition 2.2.4. The map

$$T \: (0,1) \to (0,1), \quad \alpha \mapsto \frac{1}{\alpha} - \left\lfloor \frac{1}{\alpha} \right\rfloor$$

is called the continued fraction map (or Gauss map).

Lemma 2.2.5. *The partial quotients p_r, q_r satisfy the recursion:*

$$p_{r+1} = a_{r+1}p_r + p_{r-1} \quad and \quad q_{r+1} = a_{r+1}q_r + q_{r-1}. \qquad (2.2.1)$$

Proof. See [40, p. 7].

Lemma 2.2.6. *For*

$$\alpha = [0; a_1, a_2, \ldots, a_r, a_{r+1}, \ldots],$$

we have

$$T^r \alpha = [0; a_{r+1}, a_{r+2}, \ldots] \qquad (2.2.2)$$

The map T preserves the measure

$$\omega(\mathcal{E}) = \frac{1}{\log 2} \int_{\mathcal{E}} \frac{dx}{1+x}, \qquad (2.2.3)$$

i.e., $\omega(T(\mathcal{E})) = \omega(\mathcal{E})$, for all measurable sets $\mathcal{E} \subset (0,1)$.

Proof. The result (2.2.2) is well known and can be easily confirmed by direct computation. For (2.2.3) see [38, p. 119]. □

Lemma 2.2.7. *There is a constant $A_0 > 1$, such that $q_r \geq A_0^r$, for all $r \in \mathbb{N}$.*

Proof. This is well known and easily follows from (2.2.1) of Lemma 2.2.5. □

Definition 2.2.8. Let $\alpha \in (0,1) \setminus \mathbb{Q}$, $r \in \mathbb{N}$. Then, we set

$$c(\alpha, r) = \sum_{j=0}^{r} \frac{\log q_{j+1}}{q_j},$$

$$c(\alpha, +\infty) = \sum_{j=0}^{+\infty} \frac{\log q_{j+1}}{q_j} \in \mathbb{R} \cup \{+\infty\}$$

We define the constant $c_0 > 0$ by

$$c_0 \sum_{r \geq 0} A_0^{-r/2} = \frac{1}{4}$$

and define the sequence $(w^{(r)})$ by

$$w^{(r)} = \frac{1}{2} + c_0 \sum_{j=0}^{r} A_0^{-j/2}.$$

For $z \in (0, +\infty)$, we define

$$\mathcal{E}(z, 0) := \{\alpha \in (0, 1) \setminus \mathbb{Q} : c(\alpha, 1) \geq w^{(0)} z\}, \ (w^{(0)} = 1/2),$$
$$\mathcal{E}(z, r) := \{\alpha \in (0, 1) \setminus \mathbb{Q} : c(\alpha, r-1) < w^{(r-1)} z, \ c(\alpha, r) \geq w^{(r)} z\},$$
$$\mathcal{E}(z, +\infty) := \{\alpha \in (0, +\infty) \setminus \mathbb{Q} : c(\alpha, +\infty) \geq z\}.$$

Lemma 2.2.9. *For $z \in (0, +\infty)$, we have*

$$\mathrm{meas}(\mathcal{E}(z, +\infty)) \leq \sum_{r \geq 0} \mathrm{meas}(\mathcal{E}(z, r)),$$

where meas stands for the Lebesgue measure.

Proof. Assume that $\alpha \notin \mathcal{E}(z, r)$, for every $r \in \mathbb{N} \cup \{0\}$. Then it follows by induction on r that $c(\alpha, r) \leq w^{(r)} z$ and thus by the definition of $w^{(r)}$ and the constant c_0 we have

$$c(\alpha, +\infty) = \lim_{r \to +\infty} c(\alpha, r) \leq \frac{3}{4} z.$$

Therefore, if $\alpha \in \mathcal{E}(z, +\infty)$ we have $\alpha \in \mathcal{E}(z, r)$ for at least one value of $r \in \mathbb{N} \cup \{0\}$. Thus

$$\mathcal{E}(z, +\infty) \subset \bigcup_{r=0}^{+\infty} \mathcal{E}(z, r),$$

which proves Lemma 2.2.9 $\qquad\qquad\qquad\qquad\qquad\qquad\qquad\qquad\square$

Lemma 2.2.10. *There are absolute constants $z_0 > 0$ and $c_0 > 0$, such that*

$$\mathrm{meas}(\mathcal{E}(z, r)) \leq \exp\left(-\frac{1}{2} c_0 A_0^{r/2} z\right), \quad \text{for all} \ z \geq z_0.$$

Proof. Assume that $\alpha \in \mathcal{E}(z, r)$. We have

$$c(\alpha, r) = c(\alpha, r - 1) + \frac{\log q_{r+1}}{q_r}.$$

The inequalities

$$c(\alpha, r - 1) < w^{(r-1)} z \quad \text{and} \quad c(\alpha, r) \geq w^{(r)} z$$

imply that

$$\frac{\log q_{r+1}}{q_r} \geq \left(w^{(r)} - w^{(r-1)} \right) z = c_0 A_0^{-r/2} z \qquad (2.2.4)$$

and

$$q_{r+1} \geq \exp \left(c_0 A_0^{-r/2} q_r z \right) \geq \exp \left(c_0 q_r^{1/2} z \right). \qquad (2.2.5)$$

From

$$q_{r+1} = a_{r+1} q_r + q_{r-1} \leq (a_{r+1} + 1) q_r,$$

we obtain

$$a_{r+1} \geq q_{r+1} q_r^{-1} - 1 \geq \exp \left(c_0 q_r^{1/2} z \right) q_r^{-1} - 1$$
$$\geq \exp \left(\frac{3}{4} c_0 q_r^{1/2} z \right) \geq \exp \left(\frac{3}{4} c_0 A_0^{r/2} z \right), \qquad (2.2.6)$$

if z_0 is sufficiently large. We have for all $w > 0$:

$$T^r \{ \alpha = [0; a_1, \ldots, a_{r+1}, \ldots], a_{r+1} \geq w \} = \{ \alpha = [0; a_{r+1}, \ldots], a_{r+1} \geq w \},$$

by Lemma 2.2.6. Since T preserves the measure ω, we have

$$\omega \{ \alpha = [0; a_1, \ldots, a_{r+1}, \ldots], a_{r+1} \geq w \} = \omega \{ \alpha = [0; a_{r+1}, \ldots], a_{r+1} \geq w \}.$$

Therefore $[0; a_{r+1}, \ldots] \leq w^{-1}$ and thus

$$\omega \{ \alpha = [0; a_1, \ldots, a_{r+1}, \ldots], a_{r+1} \geq w \} \leq \frac{1}{\log 2} \int_0^{w^{-1}} \frac{dx}{1+x} \leq 2w^{-1}. \qquad (2.2.7)$$

We set in (2.2.7):

$$w = \exp \left(\frac{3}{4} c_0 A_0^{r/2} z \right)$$

and obtain from (2.2.6) and (2.2.7) that

$$\text{meas}(\mathcal{E}(z,r)) \leq \exp\left(-\frac{1}{2}c_0 A_0^{r/2} z\right).$$ □

Lemma 2.2.11. *There is a constant $c_1 > 0$ such that*

$$\text{meas}(\mathcal{E}(z,+\infty)) \leq \exp(-c_1 z), \quad \text{if } z \geq z_0.$$

Proof.   This follows from Lemmas 2.2.9 and 2.2.10.   □

2.2.2. Results of R. de la Bretèche and G. Tenenbaum

R. de la Bretèche and G. Tenenbaum [23, Théorème 4.4] proved the following result.

Theorem 2.2.12. *The function*

$$g(\alpha) = \sum_{l \geq 1} \frac{1 - 2\{l\alpha\}}{l}$$

converges for $\alpha \in \mathbb{Q}$ if and only if

$$\sum_{r \geq 1} (-1)^r \frac{\log q_{r+1}}{q_r}$$

converges. In this case

$$g(\alpha) = -\sum_{m \geq 1} \frac{d(m)}{\pi m} \sin(2\pi m\alpha), \qquad (**)$$

where d stands for the divisor function.

The following definitions are adopted from [23, p. 8].

Definition 2.2.13.   For a multiplicative function g and x, y with $1 \leq y \leq x$ and $\theta \in \mathbb{R}$ we denote by

$$Z_g(x, y; \theta) := \sum_{n \in S(x,y)} g(n) \sin(2\pi\theta n),$$

where $S(x, y) = \{n \leq x : P(n) \leq y\}$, $P(n)$ being the largest prime factor of n.

We set

$$\mu(\theta; Q) := \min_{1 \leq m \leq Q} \|m\theta\| \leq \frac{1}{Q}$$

and

$$q(\theta; Q) := \min\{q : 1 \leq q \leq Q, \text{ with } \|q\theta\| = \mu(\theta; Q)\},$$

where $\| \cdot \|$ denotes the distance to the nearest integer.

We have:

Lemma 2.2.14. *Let $A > 0$. For $x \geq 2$,*

$$Q_x := \frac{x}{(\log x)^{4A+24}},$$

$$q := q(\theta; Q_x), \ a \in \mathbb{Z}, \ (a, q) = 1, |q\theta - a| \leq \frac{1}{Q_x}, \ \theta_q := \theta - \frac{a}{q}, \ \theta \in \mathbb{R},$$

one has uniformly

$$Z_d(x, x; \theta)$$
$$= x(\log x) \left\{ \frac{\sin^2(\pi\theta_q x)}{\pi q \theta_q x} + O\left(\frac{(\log q) \log(1 + (\theta_q x)^2)}{q|\theta_q|x \log x} \right) + \frac{1}{(\log x)^A} \right\}.$$

Proof. This is Lemma 11.2 of [23, pp. 64–65]. □

Definition 2.2.15. For $\theta \in \mathbb{R} \setminus \mathbb{Q}$ let

$$(q_m)_{m \geq 1} = (q_m(\theta))_{m \geq 1}$$

denote the sequence of the denominators of the partial fractions of θ. Let a_m/q_m denote the mth partial fraction of θ.
We set

$$\varepsilon_m := \theta - \frac{a_m}{q_m}.$$

The set of all real numbers for which $q(\theta; Q_x) = q_m$ is an interval defined by the conditions

$$q_m \leq Q_x < q_{m+1}.$$

We denote it by $[\xi_m, \xi_{m+1}]$.

Then, we have the following lemma.

Lemma 2.2.16. *For a positive real constant B, we have*

$$\xi_m \asymp q_m (\log q_m)^B,$$

$$|\varepsilon_m|\xi_m \asymp \frac{(\log q_m)^B}{q_{m+1}},$$

$$|\varepsilon_m|\xi_{m+1} \asymp \frac{(\log q_{m+1})^B}{q_m},$$

where $K \asymp L$ denotes $K = O(L)$ and $L = O(K)$.

Proof. This is equation (6.3) of [23, p. 22]. □

Lemma 2.2.17. *Let $\alpha \in (0,1) \setminus \mathbb{Q}$. There are constants $c_2, c_3 > 0$ such that*

$$|g(\alpha)| \le c_2 c(\alpha, +\infty) + c_3.$$

Proof. We closely follow [23, p. 65]. By partial summation, we obtain

$$g(\alpha) = \sum_{n \ge 1} \frac{d(n)}{n} \sin(2\pi n \alpha)$$

$$= \int_1^{+\infty} Z_d(t, t; \alpha) \frac{dt}{t^2}$$

$$= \sum_{m \ge 1} \left(\int_{\xi_m}^{\xi_{m+1}} Z_d(t, t; \alpha) \frac{dt}{t^2} \right).$$

By [23, equation (11.5), p. 65], we have

$$\int_{\xi_m}^{\xi_{m+1}} Z_d(t, t; \alpha) \frac{dt}{t^2} = \frac{1}{2}\pi \operatorname{sgn}(\varepsilon_m) \frac{\log q_{m+1}}{q_m} + O\left(\frac{1}{q_m^{1-1/B}} + \int_{\xi_m}^{\xi_{m+1}} \frac{dt}{t(\log t)^A} \right),$$

where A is fixed, but arbitrarily large. Therefore

$$g(\alpha) = \int_1^{+\infty} Z_d(t, t; \alpha) \frac{dt}{t^2}$$

$$\le c_2 \sum_{m \ge 1} \frac{\log q_{m+1}}{q_m} + \sum_{m \ge 1} q_m^{1-1/B} + \int_1^{+\infty} \frac{dt}{t(\log t)^A}$$

$$\le c_2 \, c(\alpha, +\infty) + c_3,$$

since the sequence $(q_m)_{m \ge 1}$ is growing exponentially and the integral converges if $A > 1$. This completes the proof. □

Proof of Theorem 2.2.1. Let $L \in \mathbb{N}$ and assume that α satisfies (**) (Théorème 4.4. of [23]) and $|g(\alpha)| \geq 4L$.

We set $z = c_2 c(\alpha, +\infty) + c_3$ and obtain by Lemmas 2.2.11 and 2.2.17 and by the definition of $c(\alpha, +\infty)$ that

$$\text{meas}\{\alpha \ : \ |g(\alpha)| \geq yL\} \leq \exp(-c_1 yL).$$

Therefore, by partitioning the set of α-values we obtain

$$\int_0^1 |g(\alpha)|^L d\alpha \leq \sum_{j \geq 0} \left((2^{j+1}L)^L \text{ meas}\{\alpha \ : \ 2^j L \leq |g(\alpha)| \leq 2^{j+1}L\} \right)$$

$$\leq \sum_{j \geq 0} (2^{j+1}L)^L \exp(-c_1 2^j L) \leq C_0^L L^L. \qquad \square$$

However,

$$H_k = \int_0^1 \left(\frac{g(x)}{2\pi} \right)^{2k} dx$$

$$= (2\pi)^{-2k} \int_0^1 g(x)^{2k} dx$$

$$\leq (2\pi)^{-2k} C_0^{2k} (2k)^{2k}$$

$$= \left(\frac{C_0}{2\pi} \right)^{2k} (2k)^{2k},$$

because of Theorem 2.2.1 with $L = 2k$, $k \in \mathbb{N}$. Also,

$$(2k)^{2k} \leq (2k)! \, 3^{2k},$$

for $k \geq k_0$, for some $k_0 \in \mathbb{N}$. Hence

$$\frac{H_k}{(2k)!} \leq \left(\frac{C_0}{2\pi} \right)^{2k} 3^{2k} = \left(\frac{3 \, C_0}{2\pi} \right)^{2k},$$

for $k \geq k_0$, for some $k_0 \in \mathbb{N}$.

Hence, the radius of convergence of the series

$$\sum_{k \geq 0} \frac{H_k}{(2k)!} x^k$$

is positive.

For the proof of Theorem 2.2.2, the following definitions and lemmas will be used. We show that the subinterval $I(k) = [0, e^{-2k}]$ gives a sufficiently big contribution to the integral

$$H_k = \int_0^1 \left(\frac{g(x)}{2\pi} \right)^{2k} dx.$$

For the details, see [60].

2.3. The Order of Magnitude for Moments for Certain Cotangent Sums

In the paper [61] Maier and Rassias sharpened the results on the moments H_k of the paper [60]. They proved the following theorem.

Theorem 2.3.1. *There are constants $c_1, c_2 > 0$, such that*

$$c_1 \Gamma(2k + 1) \leq \int_0^1 g(x)^{2k} \, dx \leq c_2 \Gamma(2k + 1),$$

for all $k \in \mathbb{N}$, where $\Gamma(\cdot)$ stands for the Gamma function.

Like in the paper [60], continued fractions are of importance. Furthermore, ideas of Balazard and Martin [11, 12], as well as ideas from the paper of Marmi, Moussa and Yoccoz [72] on Dynamical Systems play a crucial role. For the function $g(x)$ one obtains the representation

$$g(x) = \mathcal{W}(x) + H(x),$$

where $\mathcal{W}(x)$ is Wilton's function and $H(x)$ is bounded. Wilton's function is given by the following series of definitions. Let $X = (0, 1) \setminus \mathbb{Q}$.

Definition 2.3.2. Let $\alpha(x) = \{1/x\}$ for $x \in X$. The iterates α_k of α are defined by $\alpha_0(x) = x$ and

$$\alpha_k(x) = \alpha(\alpha_{k-1}(x)), \quad \text{for } k > 1.$$

Lemma 2.3.3. *Let $x \in X$ and let*

$$x = [a_0(x); a_1(x), \ldots, a_k(x), \ldots]$$

be the continued fraction expansion of x. We define the partial quotient of $p_k(x)$, $q_k(x)$ by

$$\frac{p_k(x)}{q_k(x)} = [a_0(x); a_1(x), \ldots, a_k(x)], \quad \text{where } (p_k(x), q_k(x)) = 1.$$

Then we have

$$a_k(x) = \left\lfloor \frac{1}{\alpha_{k-1}(x)} \right\rfloor,$$

$$p_{k+1} = a_{k+1}p_k + p_{k-1} \quad \text{and} \quad q_{k+1} = a_{k+1}q_k + q_{k-1}.$$

Proof.   Cf. [40, p. 7]. $\qquad\qquad\qquad\qquad\qquad\qquad\qquad\qquad\qquad\square$

Definition 2.3.4. Let $x \in X$. Let also

$$\beta_k(x) = \alpha_0(x)\alpha_1(x) \cdots \alpha_k(x)$$

(by convention $\beta_{-1} = 1$) and

$$\gamma_k(x) = \beta_{k-1}(x) \log \frac{1}{\alpha_k(x)}, \quad \text{where } k \geq 0,$$

so that

$$\gamma_0(x) = \log\left(\frac{1}{x}\right).$$

The number x is called a Wilton number, if the series

$$\sum_{k \geq 0} (-1)^k \gamma_k(x)$$

converges.
Wilton's function $\mathcal{W}(x)$ is defined by

$$\mathcal{W}(x) = \sum_{k \geq 0} (-1)^k \gamma_k(x)$$

for each Wilton number $x \in (0, 1)$.

An approximation for Wilton's function $\mathcal{W}(x)$ is given by

$$\mathcal{L}(x, n) = \sum_{\nu=0}^{n} (-1)^\nu (T^\nu l)(x),$$

where

$$l(x) = \log(1/x), \quad n \in \mathbb{N}, \ x \in X,$$

with T given by

$$T : L^p \to L^p, \quad p > 1, \quad Tf(x) = xf(\alpha(x)).$$

The approximation $\mathcal{L}(x, n)$ as well as the operator T also play a crucial role in the rest of this chapter.

In the paper [63], Maier and Rassias replace the double inequality

$$c_1\Gamma(2k + 1) \leq \int_0^1 g(x)^{2k} \, dx \leq c_2\Gamma(2k + 1),$$

for all $k \in \mathbb{N}$, where $\Gamma(\cdot)$ stands for the Gamma function, by an asymptotic result. They proved the following theorem.

Theorem 2.3.5. *Let*

$$A = \int_0^\infty \frac{\{t\}^2}{t^2} dt$$

and $K \in \mathbb{N}$. There is an absolute constant $C > 0$, such that

$$\int_0^1 |g(x)|^K dx = 2e^{-A}\Gamma(K + 1)(1 + O(\exp(-CK))), \quad \text{for } K \to \infty.$$

The value of the constant A is

$$A = \log(2\pi) - \gamma \quad \text{(see [63])}.$$

(Thanks are due to Goubi Mouloud for the information about the value of the constant). The key to the improvement of the above double inequality is the use of more subtle properties of the function H in the approximation

$$g(x) = \mathcal{W}(x) + H(x).$$

From the paper [63] we recall the following definitions and results.

Definition 2.3.6. For $\lambda \geq 0$, we set

$$A(\lambda) := \int_0^\infty \{t\}\{\lambda t\} \frac{dt}{t^2},$$

$$F(x) := \frac{x + 1}{2} A(1) - A(x) - \frac{x}{2} \log x,$$

$$G(x) := \sum_{j \geq 0} (-1)^j \beta_{j-1}(x) F(\alpha_j(x)),$$

$B_1(t) := t - \lfloor t \rfloor - 1/2$, the first Bernoulli function,

$B_2(t) := \{t\}^2 - \{t\} + 1/6$, $(t \in \mathbb{R})$ the second Bernoulli function.

For $\lambda \in \mathbb{R}$, let

$$\phi_2(\lambda) := \sum_{n \geq 1} \frac{B_2(n\lambda)}{n^2}.$$

Lemma 2.3.7. *It holds*

$$A(\lambda) = \frac{\lambda}{2} \log \frac{1}{\lambda} + \frac{1 + A(1)}{2} \lambda + O(\lambda^2), \quad as \; \lambda \to 0.$$

Proof. By [12, Proposition 31, formula (74)], we have

$$A(\lambda) = \frac{\lambda}{2} \log \frac{1}{\lambda} + \frac{1 + A(1)}{2} \lambda + \frac{\lambda^2}{2} \phi_2 \left(\frac{1}{\lambda} \right) - \int_{1/\lambda}^{\infty} \phi_2(t) \frac{dt}{t^3}.$$

From Definition 2.3.6, it follows that $\phi_2(t)$ is bounded. Therefore

$$\frac{\lambda^2}{2} \phi_2 \left(\frac{1}{\lambda} \right) = O(\lambda^2)$$

and

$$\int_{1/\lambda}^{\infty} \phi_2(t) \frac{dt}{t^3} = O(\lambda^2). \qquad \square$$

Lemma 2.3.8. *We have*

$$H(x) = 2 \sum_{j \geq 0} (-1)^{j-1} \beta_{j-1}(x) F(\alpha_j(x)).$$

In the paper [64], Maier and Rassias extend the result of [63] for arbitrary positive real values of the exponent k. They proved the following theorem.

Theorem 2.3.9. *Let $K \in \mathbb{R}$, $K > 0$. There is an absolute constant $C > 0$, such that*

$$\int_0^1 |g(x)|^K dx = \frac{e^\gamma}{\pi} \Gamma(K+1)(1 + O(\exp(-CK))),$$

for $K \to \infty$, where γ is the Euler–Mascheroni constant.

The idea behind the proof of this theorem is the replacement of the binomial theorem, which has been used for the evaluation of Kth powers in the paper [63] by the binomial series, which is an infinite series. In this case a subtle estimate of the error term occurring in the approximation of these series by a polynomial is needed.

2.4. The Distribution of Cotangent Sums for Arguments from Special Sequences and Joint Distribution for Various Arguments

In the main result of the paper [69], Maier and Rassias replaced the set of the rational numbers by the set of rational numbers p/q, where q is a fixed prime number and p runs through the primes p with

$$A_0 q \le p \le A_1 q.$$

They proved the following theorem.

Theorem 2.4.1. *With Definition 2.1.4 and the notations of Theorem 2.1.5 we have:*

For all $f \in C_0(\mathbb{R})$, the following holds true:

$$\lim_{\substack{q \to +\infty \\ q \text{ prime}}} \frac{\log q}{q} \sum_{\substack{p\,:\,A_0 q \le p \le A_1 q \\ p \text{ prime}}} f\left(\frac{1}{q} c_0\left(\frac{p}{q}\right)\right) = \int f \, d\mu.$$

2.4.1. Outline of the proof

In the sequel, the letter p will always denote a prime number. The proof is closely related to the proof of [62, Theorem 1.5].

In a first step Maier and Rassias relate $c_0(p/q)$ to the sum

$$Q\left(\frac{p}{q}\right) := \sum_{p=1}^{q-1} \cot\left(\frac{\pi m p}{q}\right) \left\lfloor \frac{mp}{q} \right\rfloor.$$

By [62, Proposition 1.8] (see also the second author's PhD thesis [94]), they proved

$$c_0\left(\frac{p}{q}\right) = \frac{1}{p} c_0\left(\frac{1}{q}\right) - \frac{1}{p} Q\left(\frac{p}{q}\right).$$

Thus — like in [62, 95] — the investigation of moments of c_0 can be reduced to the study of moments of Q. This time the sum of powers of $Q(p/q)$ in the definition of the moments is extended over primes only. In the present work, they replaced the summation over

$$\{r : (r, b) = 1,\ A_0 b \le r \le A_1 b\}$$

by the summation over $\{p \,:\, A_0 q \le p \le A_1 q\}$. The range of summation is split into subintervals in which $\left\lfloor \frac{pm}{q} \right\rfloor$ assumes constant values.

Definition 2.4.2. For $j \in \mathbb{N}$ we set

$$S_j := \{pm \ : \ qj \leq pm \leq q(j+1), \ m \in \mathbb{Z}\}$$

and write

$$S_j =: \{qj + s_j, \ qj + s_j + p, \dots, \ qj + s_j + d_j p\}.$$

We also define t_j by

$$qj + s_j + d_j p + t_j = q(j+1).$$

Lemma 2.4.3. *We have* $d_j \in \{0,1\}$. *The set* S_j *is of one of the two forms:*

$$S_j = \{qj + s_j\} \quad or \quad S_j = \{qj + s_j, \ qj + s_j + p\}.$$

Proof. This follows because of $1/2 < A_0 < A_1 < 1$. $\qquad\square$

Of crucial importance is the map $j \longrightarrow s_j$ and its inverse $s \longrightarrow j(s)$. The following congruences hold true:

$$s_j \equiv -qj \ (\mathrm{mod}\ p)$$

and

$$t_j \equiv q(j+1) \ (\mathrm{mod}\ p).$$

Because of the poles of the function $\cot(\pi x)$ at $x = 0$ and $x = 1$, small values of s_j and t_j will dominate the sum $Q(p/q)$. For the other values of s_j and t_j we expect cancellation because of the antisymmetry of the function $\cot(\pi x)$:

$$\cot(\pi(1-x)) = -\cot(\pi x).$$

To guarantee this cancellation, it is necessary to localize p^*/q and p/q simultaneously.

Whereas in the analogous problem in [62] this was achieved by the use of Kloosterman sums, they here needed a result on character sums in finite fields over rational functions of primes due to Fouvry and Michel [37]. This result will be described in Section 3.

The quality of this cancellation will depend on good equidistribution properties of the fractions $jq/p(\mathrm{mod}\ 1)$. It is well known that the equidistribution of $jq/p(\mathrm{mod}\ 1)$ is good, if p/q has no good Diophantine approximation by fractions with small denominators. These questions will be discussed in Section 4.

They determined the cardinality of exceptional sets $\mathcal{E}(m)$, consisting of prime numbers p, for which there exists an extraordinarily good approximation which is determined by the index m.

In Section 6, they carried out a decomposition $Q = Q_0 + Q_1$, where the dominating part Q_0 belongs to small values of s_j and t_j and the remainder Q_1.

In Section 7, they compared Q_0 with a partial sum of $g(\alpha)$ and show that the difference is small.

In Section 8, they then decomposed the remainder Q_1 into a sum

$$Q_1 = Q^{(1)} + Q^{(2)} + Q^{(3)},$$

where $Q^{(i)}$ is defined by the form of S_j. They then used the equidistribution properties from Section 4 to estimate the $Q^{(i)}$ and thus Q_1.

Based on the results of Sections 7 and 8, in Sections 9 and 10 they proved results on moments of $Q(p/q)$ and $c_0(p/q)$, using standard analytical tools, like Hölder's inequality.

Finally, in Section 11 they used the Riesz representation theorem to complete the proof of Theorem 2.1.5.

2.4.2. Exponential sums over primes in finite fields

Lemma 2.4.4. *Let $\mathbb{F}_r$ be the finite field with r elements and ψ be a nontrivial additive character over $\mathbb{F}_r$, f a rational function of the form*

$$f(x) = \frac{P(x)}{Q(x)},$$

P and Q relatively prime monic polynomials,

$$S(f; r, x) := \sum_{p \leq x} \psi(f(p))$$

(p denotes the p-fold sum of the element 1 in $\mathbb{F}_r$).
Then, we have

$$S(f; r, x) \ll r^{3/16+\epsilon} x^{25/32}.$$

The implied constant depends only on ϵ and the degrees of P and Q.

Proof. This is due to Fouvry and Michel [37]. $\qquad\qquad\square$

The major new idea here is contained in Lemma 2.4.5, which we will now discuss in detail.

Lemma 2.4.5. *Let $1/2 < A_0 < A_1 < 1$, and $r \in \mathbb{N}$. Let $\alpha \in (0,1)$, $\delta > 0$, such that $\alpha + \delta < 1$. We denote*

$$q^* := q^*(p,q) \ by \ qq^* \equiv 1 \ (\mathrm{mod} \ p)$$

and

$$p^* := p^*(p,q) \in \mathbb{N} \ by \ pp^* \equiv 1 \ (\mathrm{mod} \ q).$$

Then, we have

$$N(\alpha, \delta) := \left| \left\{ p \ : \ p \ prime, \ A_0 q \le p \le A_1 q, \ \alpha \le \frac{q^*}{p} \le \alpha + \delta \right\} \right|$$
$$= \delta(A_1 - A_0) \frac{q}{\log q} (1 + o(1)), \quad (q \to +\infty).$$

Proof. The Diophantine equation $qx + py = 1$ has exactly one solution (x_0, y_0) with

$$-\left\lfloor \frac{p}{2} \right\rfloor < x_0 \le \left\lfloor \frac{p}{2} \right\rfloor, \quad -\left\lfloor \frac{q}{2} \right\rfloor < y_0 \le \left\lfloor \frac{q}{2} \right\rfloor.$$

We have

$$q^* \equiv x_0 \ (\mathrm{mod} \ p), \ p^* \equiv y_0 \ (\mathrm{mod} \ q). \tag{2.4.1}$$

Therefore, for

$$\beta \in (-1/2, 1/2) \text{ and } \delta > 0 \text{ with } \beta + \delta < 1/2 \text{ and } \beta - \delta > -1/2$$

we have

$$\left| \left\{ p \ : \ p \ prime, \ A_0 q \le p \le A_1 q, \ \frac{y_0}{q} \in [\beta, \beta + \delta] \right\} \right|$$
$$= \left| \left\{ p \ : \ p \ prime, \ A_0 q \le p \le A_1 q, \ \frac{x_0}{p} \in [-(\beta + \delta), -\beta] \right\} \right| + O(1)$$
$$= \left| \left\{ p \ : \ p \ prime, \ A_0 q \le p \le A_1 q, \ \frac{q^*}{p} (\mathrm{mod} \ 1) \in [-(\beta + \delta), -\beta] \right\} \right| + O(1),$$
$$\tag{2.4.2}$$

where

$$\frac{q^*}{p} \ (\mathrm{mod} \ 1) \in [-(\beta + \delta), -\beta]$$

stands for

$$\frac{q^*}{p} \in \begin{cases} [-(\beta+\delta), -\beta] + 1 & \text{if } \beta \geq 0, \\ [-(\beta+\delta), -\beta] & \text{if } \beta < 0. \end{cases}$$

Let $\Delta > 0$ such that

$$\beta + \delta + \Delta \leq 1/2, \ \ 0 \leq v \leq \Delta.$$

We define the functions

$$\chi_1(u, v) := \begin{cases} 1 & \text{if } u \in [\beta + \Delta - v, \beta + \delta - \Delta + v], \\ 0 & \text{otherwise.} \end{cases} \tag{2.4.3}$$

and

$$\chi_2(u, v) := \begin{cases} 1 & \text{if } u \in [\beta - \Delta - v, \beta + \delta - \Delta - v], \\ 0 & \text{otherwise.} \end{cases} \tag{2.4.4}$$

as well as the functions l_1, l_2 by

$$l_i(u) := \Delta^{-1} \int_0^{\Delta} \chi_i(u, v) \, dv \quad \text{for } i = 1, 2.$$

We define the function

$$\tilde{\chi}(p, \beta) := \begin{cases} 1 & \text{if } \frac{p^*}{q} \in [\beta, \beta + \delta], \\ 0 & \text{otherwise.} \end{cases} \tag{2.4.5}$$

Since l_i for $i = 1, 2$, is obtained from χ_i by averaging over v and since

$$0 \leq \chi_i(u, v) \leq 1 \quad \text{for } i = 1, 2,$$

it follows that

$$0 \leq l_i(u) \leq 1 \quad \text{for } i = 1, 2.$$

We have

$$l_1\left(\frac{p^*}{q}\right) = 0, \text{ if } \frac{p^*}{q} \notin [\beta, \beta + \delta).$$

Similarly, we obtain

$$l_2\left(\frac{p^*}{q}\right) = 1, \text{ if } \frac{p^*}{q} \in [\beta, \beta + \delta).$$

Thus we obtain

$$l_1\left(\frac{p^*}{q}\right) \leq \tilde{\chi}(p,\beta) \leq l_2\left(\frac{p^*}{q}\right). \tag{2.4.6}$$

By a computation as in [62, Lemma 4.3], we have the Fourier expansion

$$l_i(u) = \sum_{n=-\infty}^{+\infty} a(n)e(nu), \quad \text{for } i = 1,2,$$

where

$$a(0) = \delta + R_1, \quad |R_1| \leq \Delta$$

and

$$a(n) = \begin{cases} O(\Delta) & \text{if } |n| \leq \Delta^{-1}, \\ O(\Delta^{-1}n^2) & \text{if } |n| > \Delta^{-1}. \end{cases} \tag{2.4.7}$$

Let $\Delta_1 > 0$ such that

$$A_0 - \Delta_1 > 1/2, \quad A_1 + \Delta_1 < 1 \text{ and } 0 \leq v \leq \Delta_1.$$

We define the functions

$$\chi_3(u,v) := \begin{cases} 1 & \text{if } u \in [A_0 + v - \Delta_1, A_1 - v + \Delta_1], \\ 0 & \text{otherwise} \end{cases} \tag{2.4.8}$$

and

$$\chi_4(u,v) := \begin{cases} 1 & \text{if } u \in [A_0 + \Delta_1 - v, A_1 + \Delta_1 + v], \\ 0 & \text{otherwise} \end{cases} \tag{2.4.9}$$

as well as the functions l_3, l_4 by

$$l_i(u) := \Delta_1^{-1} \int_0^{\Delta_1} \chi_i(u,v)\, dv \quad \text{for } i = 3,4.$$

We define the function

$$\chi^*(p,\beta) := \begin{cases} 1 & \text{if } A_0 \leq \frac{p}{q} \leq A_1, \\ 0 & \text{otherwise.} \end{cases} \tag{2.4.10}$$

Since l_i, for $i = 3,4$, is obtained from χ_i by averaging over v and since

$$0 \leq \chi_i(u,v) \leq 1 \text{ for } i = 3,4,$$

we obtain

$$0 \le l_i(u) \le 1 \quad \text{for } i = 3, 4.$$

From (2.4.8) we have

$$l_3\left(\frac{p}{q}\right) = 0, \quad \text{if } \frac{p}{q} \notin (A_0, A_1).$$

From (2.4.9) we have

$$l_3\left(\frac{p}{q}\right) = 1, \quad \text{if } \frac{p}{q} \in (A_0, A_1).$$

Therefore, we obtain

$$l_3\left(\frac{p}{q}\right) \le \chi^*(p, \beta) \le l_4\left(\frac{p}{q}\right). \tag{2.4.11}$$

An analogous computation as for l_1, l_2 gives the Fourier expansions

$$l_i(u) = \sum_{n=-\infty}^{+\infty} c(n)e(nu), \quad \text{for } i = 3, 4$$

with

$$c(0) = A_1 - A_0 + R_2, \quad \text{where } |R_2| \le \Delta_1$$

and

$$c(n) = \begin{cases} O(1) & \text{if } |n| \le \Delta_1^{-1}, \\ O(\Delta_1^{-1}n^{-2}) & \text{if } |n| > \Delta_1^{-1}. \end{cases} \tag{2.4.12}$$

From (2.4.2), (2.4.5), (2.4.6), (2.4.10) and (2.4.11), setting $\beta = -\alpha$, we get the following expression:

$$\sum_{1 \le p \le q-1} l_1\left(\frac{p^*}{q}\right) l_3\left(\frac{p}{q}\right) \le N(\alpha, \delta) \le \sum_{1 \le p \le q-1} l_2\left(\frac{p^*}{q}\right) l_4\left(\frac{p}{q}\right). \tag{2.4.13}$$

Therefore

$$\sum_{1 \le p \le q-1} l_1\left(\frac{p^*}{q}\right) l_3\left(\frac{p}{q}\right) = \sum_{m,n=-\infty}^{+\infty} a(m)c(n)E(n, m, q) \tag{2.4.14}$$

with

$$E(n, m, q) := \sum_{1 \le p \le q-1} e\left(\frac{np + mp^*}{q}\right).$$

For $(m, n) \neq (0, 0)$ we again estimate $E(n, m, q)$ by application of Lemma 2.5.3, where we set

$$r = q, \ P(x) = x^2 + 1, \ Q(x) = x, \psi(l \bmod \ q) = e\left(\frac{nl}{q}\right)$$

and obtain

$$E(m, n) \ll q^{31/32}.$$

We obtain from (2.4.14) the following expression:

$$\sum_{1 \leq p \leq q-1} l_1\left(\frac{p^*}{q}\right) l_3\left(\frac{p}{q}\right) = (\delta + R_1)(A_1 - A_0 + R_2)\frac{q}{\log q} + o\left(\frac{q}{\log q}\right)$$

$$(2.4.15)$$

for $|R_1| \leq \Delta$ and $|R_2| \leq \Delta$. By the same computation we also get

$$\sum_{1 \leq p \leq q-1} l_2\left(\frac{p^*}{q}\right) l_4\left(\frac{p}{q}\right) = (\delta + R_1)(A_1 - A_0 + R_2)\frac{q}{\log q} + o\left(\frac{q}{\log q}\right).$$

$$(2.4.16)$$

Therefore

$$\sum_{1 \leq p \leq q-1} l_2\left(\frac{p^*}{q}\right) l_4\left(\frac{p}{q}\right) = \delta(A_1 - A_0)\frac{q}{\log q} + \delta R_2\frac{q}{\log q} + R_1(A_1 - A_0)\frac{q}{\log q}$$

$$+ R_1 R_2 \frac{q}{\log q} + o\left(\frac{q}{\log q}\right).$$

$$(2.4.17)$$

Since Δ and Δ_1 can be chosen to be arbitrarily small, it follows that (2.4.17) implies Lemma 2.4.5. $\qquad\square$

In the paper [70], Maier and Rassias simultaneously consider the values of c_0 for various shifts of the argument. They consider general numerators as well as prime numerators. For simplicity, the denominator will be a fixed prime q. Their main results are the following.

Theorem 2.4.6. *Let A_0, A_1 be fixed constants such that $1/2 < A_0 < A_1 < 1$. Let $a_1, a_2, \ldots, a_L$ be distinct non-negative integers. Let $f_1, f_2, \ldots, f_L \in C_0(\mathbb{R})$ $(L \in \mathbb{N})$. Then*

(i)

$$\lim_{\substack{q \to +\infty \\ q \ prime}} \frac{1}{\phi(q)} \sum_{r \,:\, A_0 q \le r \le A_1 q} \frac{1}{(A_1 - A_0)^L} \prod_{l=1}^{L} f_l\left(c_0\left(\frac{r + a_l}{q}\right)\right)$$

$$= \prod_{l=1}^{L}\left(\int f_l \, d\mu\right),$$

(ii)

$$\lim_{\substack{q \to +\infty \\ q \ prime}} \frac{\log q}{q} \sum_{\substack{p \,:\, A_0 q \le p \le A_1 q \\ p \ prime}} \frac{1}{(A_1 - A_0)^L} \prod_{l=1}^{L} f_l\left(c_0\left(\frac{p + a_l}{q}\right)\right)$$

$$= \prod_{l=1}^{L}\left(\int f_l \, d\mu\right).$$

The proof of (ii) can be obtained from the proof of (i) by only minor changes. They give only a detailed proof of (i) and sketch the changes needed for the proof of (ii).

2.4.3. Outline of the proof

Several fundamental ideas already appear in the paper [62] and in the thesis [94]. The key to the treatment of the sum $c_0(r/q)$ lies in its relation to the sum

$$Q\left(\frac{r}{q}\right) := \sum_{m=1}^{q-1} \cot\left(\frac{\pi m r}{q}\right) \left\lfloor \frac{mr}{q} \right\rfloor.$$

The second author in his thesis [94] established the following (see also [62, Proposition 1.8]) expression:

$$c_0\left(\frac{r}{q}\right) = \frac{1}{r} c_0\left(\frac{1}{q}\right) - \frac{1}{r} Q\left(\frac{r}{q}\right). \tag{2.4.18}$$

They now have to consider simultaneously the values

$$f_1\left(Q\left(\frac{r + a_1}{q}\right)\right), \ldots, f_L\left(Q\left(\frac{r + a_L}{q}\right)\right).$$

By the Weierstrass approximation theorem, this question may be reduced to the study of the joint distribution of the products

$$\prod(k_1, \ldots, k_L) := Q\left(\frac{r+a_1}{q}\right)^{k_1} \cdots Q\left(\frac{r+a_L}{q}\right)^{k_L} \qquad (2.4.19)$$

for L-tuplets $(k_1, \ldots, k_L)$ of non-negative integers. In a similar fashion as in the previous papers, they break up the range of summation into subintervals in which $\left\lfloor \frac{(r+a_l)m}{q} \right\rfloor$ assumes constant values.

Definition 2.4.7. For $j \in \mathbb{N}$, $l \in \{1, \ldots, L\}$ we set:

$$S_j^{(l)} := \{(r+a_l)m \; : \; qj \le (r+a_l)m < q(j+1)\}$$

and write

$$S_j^{(l)} := \left\{qj + s_j^{(l)}, qj + s_j^{(l)} + (r+a_l), \ldots, qj + s_j^{(l)} + (r+a_l)d_j^{(l)}\right\}.$$

We also define $t_j^{(l)}$ by

$$qj + s_j^{(l)} + d_j^{(l)}(r+a_l) + t_j^{(l)} := q(j+1).$$

In [62, 69] the map $j \longrightarrow s_j$ and its inverse $s \longrightarrow j(s)$ were very important.

They are now replaced by L maps $j \longrightarrow s_j^{(l)}$ $(1 \le l \le L)$ and their inverses $s \longrightarrow j_l(s)$.

We also have L pairs of congruences

$$s_j^{(l)} \equiv -qj \, (\mathrm{mod} \, (r+a_l)) \quad \text{and} \quad t_j^{(l)} \equiv q(j+1) \, (\mathrm{mod} \, (r+a_l)). \quad (2.4.20)$$

Each of the sums $Q\left(\frac{r+a_l}{q}\right)$ is dominated by small values of $s_j^{(l)}$ and $t_j^{(l)}$ because of the poles of the function $\cot(\pi x)$ at $x = 0$ and $x = 1$. We denote this partial sum by $Q_0\left(\frac{r+a_l}{q}\right)$ and thus we obtain the decomposition

$$Q\left(\frac{r+a_l}{q}\right) = Q_0\left(\frac{r+a_l}{q}\right) + Q_1\left(\frac{r+a_l}{q}\right).$$

The function $\cot(\pi x)$ is antisymmetric

$$\cot(\pi(1-x)) = -\cot(\pi x).$$

Therefore, there will be considerable cancellation in the sums $Q_1\left(\frac{r+a_l}{q}\right)$, which thus will be small.

By the binomial theorem, the products $\prod(k_1, \ldots, k_L)$ in (2.4.18) will be linear combinations of products of the form

$$Q_{\epsilon_1}\left(\frac{r + a_{l_1}}{q}\right)^{h_1} \cdots Q_{\epsilon_M}\left(\frac{r + a_{l_M}}{q}\right)^{h_M} \tag{2.4.21}$$

with $\epsilon_g \in \{0, 1\}$. They show that only the products with $\epsilon_g = 0$ for $1 \leq g \leq M$ will give a substantial contribution.

The asymptotic size of these products will be determined by localizing the solutions of the congruences (2.4.20) simultaneously for all l. Whereas in [62] Kloosterman sums could be used for this localization, here they need results on more general exponential sums in finite fields, due to Weil [126, 127], as well as Fouvry and Michel [37].

The contribution of the other products in (2.4.21) is small, since at least one $\epsilon_g = 1$.
For the discussion of these factors

$$Q_1\left(\frac{r + a_{l_g}}{q}\right)^{h_g}$$

they refer to results of [62].

2.4.4. Exponential sums in finite fields

Lemma 2.4.8. *Let q be a prime number. Let $a_1, \ldots, a_L$ be distinct non-negative integers, $n, m_1, \ldots, m_L$ integers not all 0,*

$$R(x) = nx + \frac{m_1}{x + a_1} + \cdots + \frac{m_L}{x + a_L}.$$

Then

$$\sum_{\substack{x=1 \\ x \neq -a_l, \, 1 \leq l \leq L}}^{q-1} e\left(\frac{R(x)}{q}\right) = O(q^{1/2}).$$

The constant implied by the O-symbol may depend on L.

Proof.   This follows from work of Weil [126, 127].   □

Lemma 2.4.9. *Let $\mathbb{F}_r$ be the finite field with r elements and ψ be a non-trivial additive character over $\mathbb{F}_r$, f a rational function of the form*

$f(x) = \frac{P(x)}{Q(x)}$, *P and Q relatively prime monic non-constant polynomials,*

$$S(f; r, x) := \sum_{p \leq x} \psi(f(p))$$

(p denotes the p-fold sum of the element 1 in $\mathbb{F}_r$). Then

$$S(f; r, x) \ll r^{3/16+\epsilon} \, x^{25/32}.$$

The implied constant depends only on ϵ and the degrees of P and Q.

Proof. This is due to Fouvry and Michel [37]. $\square$

2.4.5. Localizations of the solutions of the congruences

Lemma 2.4.10. *Let q be prime. Let $1/2 < A_0 < A_1 < 1$ and $r \in \mathbb{N}$. Let $q^*(r; l)$ be defined by*

$$qq^*(r; l) \equiv 1 \,(\mathrm{mod}\,(r + a_l)).$$

Let $\alpha_1, \ldots \alpha_L \in (0, 1)$, $\delta > 0$, such that $\alpha_l + \delta < 1$ for $1 \leq l \leq L$. Then we have

$$N(\alpha_1, \ldots \alpha_L, \delta) := \left| \left\{ r \,:\, r \in \mathbb{N},\, A_0 q \leq r \leq A_1 q,\, \alpha_l \leq \frac{q^*(r; l)}{r} \leq \alpha + \delta \right\} \right|$$

$$= \delta^L (A_1 - A_0) q (1 + o(1)), \quad q \to \infty.$$

Proof. In the sequel, we assume $1 \leq l \leq L$. We let $(r + a_l)^*$ be determined by

$$(r + a_l)(r + a_l)^* \equiv 1 \,(\mathrm{mod}\, q).$$

The Diophantine equation $qx + (r + a_l)y = 1$ has exactly one solution $(x_{0,l},\, y_{0,l})$ with.

$$-\left\lfloor \frac{r + a_l}{2} \right\rfloor < x_{0,l} \leq \left\lfloor \frac{r + a_l}{2} \right\rfloor, \quad -\left\lfloor \frac{q}{2} \right\rfloor < y_{0,l} \leq \frac{q}{2}.$$

We have

$$q^*(r; l) \equiv x_{0,l} \,(\mathrm{mod}\,(r + a_l)),$$

$$(r + a_l)^* \equiv y_{0,l} \,(\mathrm{mod}\, q). \tag{2.4.22}$$

Therefore, for $\beta_l \in (-1/2, 1/2)$ and $\delta > 0$ with

$$\beta_l + \delta < \frac{1}{2} \quad \text{and} \quad \beta - \delta > -\frac{1}{2} \quad \text{for} \ \ 1 \le l \le L,$$

we have

$$\left| \left\{ r \ : \ A_0 q \le r \le A_1 q, \ \frac{y_{0,l}}{q} \in [\beta_l, \beta_l + \delta] \right\} \right|$$

$$= \left| \left\{ r \ : \ A_0 q \le r \le A_1 q, \ \frac{x_{0,l}}{r} \in [-(\beta_l + \delta), -\beta_l] \right\} \right| + O(1)$$

$$= \left| \left\{ r \ : \ A_0 q \le r \le A_1 q, \ \frac{q^*(r, l)}{r} \ (\text{mod } 1) \in [-(\beta_l + \delta), -\beta_l] \right\} \right| + O(1),$$

$$(2.4.23)$$

where

$$\frac{q^*(r, l)}{r} \ (\text{mod } 1) \in [-(\beta_l + \delta), -\beta_l]$$

stands for

$$\frac{q^*(r, l)}{r} \in \begin{cases} [-(\beta_l + \delta), -\beta_l] + 1 & \text{if } \beta_l \ge 0, \\ [-(\beta_l + \delta), -\beta_l] & \text{if } \beta_l < 0. \end{cases}$$

Let $\Delta > 0$ such that

$$\beta_l + \delta + \Delta \le \frac{1}{2}, \quad 0 \le v_l \le \Delta.$$

We define the function

$$\chi_{1,l}(u, v) := \begin{cases} 1 & \text{if } u \in [\beta_l + \Delta - v, \ \beta_l + \delta - \Delta + v), \\ 0 & \text{otherwise}, \end{cases} \qquad (2.4.24)$$

and

$$\chi_{2,l}(u, v) := \begin{cases} 1 & \text{if } u \in [\beta_l - \Delta - v, \ \beta_l + \delta - \Delta - v), \\ 0 & \text{otherwise}, \end{cases} \qquad (2.4.25)$$

as well as the functions $\lambda_{1,l}$, $\lambda_{2,l}$ by

$$\lambda_{i,l}(u) := \Delta^{-1} \int_0^\Delta \chi_{i,l}(u, v) \, dv \quad \text{for } i = 1, 2.$$

Let the function

$$\tilde{\chi}_l(r, \beta) := \begin{cases} 1 & \text{if } \frac{(r + a_l)^*}{q} \in [\beta_l, \beta_l + \delta], \\ 0 & \text{otherwise}. \end{cases} \qquad (2.4.26)$$

Since $\lambda_{i,l}$, for $i = 1, 2$, is obtained from $\chi_{i,l}$ by averaging over r and since

$$0 \leq \chi_{i,l}(u, v) \leq 1,$$

it follows that

$$0 \leq \lambda_{i,l}(u) \leq 1 \quad \text{for } i = 1, 2.$$

From (2.4.24) we have

$$\lambda_{1,l}\left(\frac{(r + a_l)^*}{q}\right) = 0 \quad \text{if} \quad \frac{(r + a_l)^*}{q} \notin [\beta_l, \beta_l + \delta).$$

Similarly, from (2.4.25) we have

$$\lambda_{2,l}\left(\frac{(r + a_l)^*}{q}\right) = 1 \quad \text{if} \quad \frac{(r + a_l)^*}{q} \notin [\beta_l, \beta_l + \delta).$$

Thus we obtain

$$\lambda_{1,l}\left(\frac{(r + a_l)^*}{q}\right) \leq \tilde{\chi}_l(r, \beta) \leq \lambda_{2,l}\left(\frac{(r + a_l)^*}{q}\right). \tag{2.4.27}$$

We have the Fourier expansion

$$\lambda_{i,l}(u) = \sum_{n=-\infty}^{\infty} a_l(n) e(nu).$$

The Fourier coefficients $a_l(n)$ are computed as follows.

For $i = 1$:

$$a_l(0) = \Delta^{-1} \int_0^\Delta \left(\int_{\beta_l - \Delta + v}^{\beta_l + \delta + \Delta - v} 1 \, du \right) dv = \delta + \Delta,$$

as well as

$$a_l(n) = \Delta^{-1} \int_0^\Delta \left(\int_{\beta_l - \Delta + v}^{\beta_l + \delta + \Delta - v} e(-nu) \, du \right) dv$$

$$= \Delta^{-1} \int_0^\Delta -\frac{1}{2\pi i n} (e(-n(\beta_l + \delta + \Delta - v)) - e(-n(\beta_l - \Delta + v))) \, dv$$

$$= -\frac{1}{4\pi^2 n^2} \Delta^{-1} (e(-n(\beta_l + \delta)) - e(-n(\beta_l + \delta + \Delta)) - e(-n\beta_l)$$

$$+ e(-n(\beta_l - \Delta))).$$

From the above and an analogous computation, for $i = 2$ we obtain $a_l(0) = \delta + R_{1,l}$, where $|R_{1,l}| \leq \Delta$ and

$$a_l(n) = \begin{cases} O(\Delta) & \text{if } |n| \leq \Delta^{-1}, \\ O(\Delta^{-1} n^2) & \text{if } |n| > \Delta^{-1}. \end{cases} \tag{2.4.28}$$

Let $\Delta_1 > 0$ such that $A_0 - \Delta_1 > 1/2$, $A_1 + \Delta_1 < 1$ and $0 \leq v \leq \Delta_1$. We define the functions

$$\chi_3(u, v) := \begin{cases} 1 & \text{if } u \in [A_0 + v - \Delta_1, A_1 - v + \Delta_1], \\ 0 & \text{otherwise,} \end{cases} \tag{2.4.29}$$

and

$$\chi_4(u, v) := \begin{cases} 1 & \text{if } u \in [A_0 + \Delta_1 - v, A_1 + \Delta_1 + v], \\ 0 & \text{otherwise,} \end{cases} \tag{2.4.30}$$

as well as the functions λ_3, λ_4 by

$$\lambda_i(u) := \Delta_1^{-1} \int_0^{\Delta_1} \chi_i(u, v)\, dv \quad \text{for } i = 3, 4.$$

We define the function

$$\chi^*(r, \beta) := \begin{cases} 1 & \text{if } A_0 \leq \frac{r}{q} \leq A_1, \\ 0 & \text{otherwise.} \end{cases} \tag{2.4.31}$$

Since λ_i for $i = 3, 4$ is obtained from χ_i by averaging over r and since

$$0 \leq \chi_i(u, v) \leq 1 \quad \text{for } i = 3, 4,$$

we obtain

$$0 \leq \lambda_i(u) \leq 1 \quad \text{for } i = 3, 4.$$

From the above and an analogous computation, for $i = 1, 2$ we obtain

$$a_l(0) = \delta + R_{1,l}, \quad \text{where } |R_{1,l}| \leq \Delta,$$

and

$$a_l(n) := \begin{cases} O(\Delta) & \text{if } |n| \leq \Delta^{-1}, \\ O(\Delta^{-1} n^2) & \text{if } |n| > \Delta^{-1}. \end{cases}$$

Let $\Delta_1 > 0$ such that $A_0 - \Delta_1 > 1/2$, $A_1 + \Delta_1 < 1$ and $0 \le v \le \Delta_1$. We define the functions

$$\chi_3(u, v) := \begin{cases} 1 & \text{if } u \in [A_0 + v - \Delta_1, A_1 - v + \Delta_1], \\ 0 & \text{otherwise,} \end{cases}$$

and

$$\chi_4(u, v) := \begin{cases} 1 & \text{if } u \in [A_0 + \Delta_1 - v, A_1 + \Delta_1 + v], \\ 0 & \text{otherwise,} \end{cases}$$

as well as the functions λ_3, λ_4 by

$$\lambda_i(u) := \Delta_1^{-1} \int_0^{\Delta_1} \chi_i(u, v) \, dv \quad \text{for } i = 3, 4.$$

We define the function

$$\chi^*(r, \beta) := \begin{cases} 1 & \text{if } A_0 \le \frac{r}{q} \le A_1, \\ 0 & \text{otherwise.} \end{cases}$$

Since λ_i, for $i = 3, 4$, is obtained from χ_i by averaging over v and since

$$0 \le \chi_i(u, v) \le 1 \quad \text{for } i = 3, 4,$$

we obtain

$$0 \le \lambda_i(u) \le 1 \quad \text{for } i = 3, 4.$$

From (2.4.29) we have

$$\lambda_3\left(\frac{r}{q}\right) = 0 \quad \text{if } \frac{r}{q} \notin (A_0, A_1).$$

From (2.4.30) we have

$$\lambda_3\left(\frac{r}{q}\right) = 1 \quad \text{if } \frac{r}{q} \in (A_0, A_1).$$

Therefore, we obtain

$$\lambda_3\left(\frac{r}{q}\right) \le \chi^*(r, \beta) \le \lambda_4\left(\frac{r}{q}\right). \tag{2.4.32}$$

By an analogous computation as for λ_1, λ_2, we obtain the Fourier expansions

$$\lambda_i(u) = \sum_{n=-\infty}^{\infty} c(n)e(nu), \quad \text{for } i = 3, 4,$$

with

$$c(0) = A_1 - A_0 + R_2, \quad \text{where } |R_2| \le \Delta_1$$

and

$$c(n) := \begin{cases} O(1) & \text{if } |n| \le \Delta_1^{-1}, \\ O(\Delta_1^{-1}n^{-2}) & \text{if } |n| > \Delta_1^{-1}. \end{cases}$$

From (2.4.23), (2.4.26), (2.4.27), (2.4.31) and (2.4.32), setting $\beta = -\alpha$, we get the following expression:

$$\sum_{1 \le r \le q-1} \left(\prod_{l=1}^{L} \lambda_{1,l} \left(\frac{(r + a_l)^*}{q} \right) \right) \lambda_3 \left(\frac{r}{q} \right)$$

$$\le N(\alpha_1, \ldots, \alpha_L, \delta) \le \sum_{1 \le r \le q-1} \left(\prod_{l=1}^{L} \lambda_{2,l} \left(\frac{(r + a_l)^*}{q} \right) \right) \lambda_4 \left(\frac{r}{q} \right).$$

We obtain

$$\sum_{1 \le r \le q-1} \left(\prod_{l=1}^{L} \lambda_{1,l} \left(\frac{(r + a_l)^*}{q} \right) \right) \lambda_3 \left(\frac{r}{q} \right)$$

$$= \sum_{m_1, \ldots, m_L, n = -\infty}^{\infty} a(m_1)a(m_2) \cdots a(m_L)c(n)E(n, m_1, \ldots, m_L, q),$$

$$(2.4.33)$$

with

$$E(n, m_1, \ldots, m_L, q) := \sum_{1 \le r \le q-1} e\left(\frac{nr + m_1(r + a_1)^* + \cdots + m_L(r + a_L)^*}{q} \right).$$

For $(n, m_1, \ldots, m_L) \ne (0, 0, \ldots, 0)$ we estimate $E(n, m_1, \ldots, m_L, q)$ by Lemma 2.4.8 and obtain

$$E(n, m_1, \ldots, m_L, q) = O(q^{1/2}).$$

From (2.4.33) we get

$$\sum_{1\leq r\leq q-1}\left(\prod_{l=1}^{L}\lambda_{1,l}\left(\frac{(r+a_l)^*}{q}\right)\right)\lambda_3\left(\frac{r}{q}\right) = (\delta + R_1)^L(A_1 - A_0 + R_2)q + o(q),$$

(2.4.34)

for $|R_1| \leq \Delta$ and $|R_2| \leq \Delta$. The same computation also gives

$$\sum_{1\leq r\leq q-1}\left(\prod_{l=1}^{L}\lambda_{2,l}\left(\frac{(r+a_l)^*}{q}\right)\right)\lambda_4\left(\frac{r}{q}\right) = (\delta + R_1)^L(A_1 - A_0 + R_2)q + o(q).$$

(2.4.35)

Since Δ and Δ_1 can be chosen to be arbitrarily small, it follows that (2.4.34) and (2.4.35) imply Lemma 2.4.10. □

2.4.6. Decomposition of the sums Q

Maier and Rassias use decompositions of the sums Q, which are similar to the decompositions used in [62].

$$Q\left(\frac{r+a_l}{q}\right) = \sum_{j=1}^{r-1} j \sum_{h=0}^{d_j} \cot\left(\pi \frac{s_j^{(l)} + hr}{q}\right).$$

We further decompose $Q\left(\frac{r+a_l}{q}\right)$ as in the following definition.

Definition 2.4.11.

$$Q\left(\frac{r+a_l}{q}\right) := Q_0\left(\frac{r+a_l}{q}\right) + Q_1\left(\frac{r+a_l}{q}\right)$$

with

$$Q_0\left(\frac{r+a_l}{q}\right) := \sum_{j=1}^{q-1}{}^* j \sum_{h=0}^{d_j^{(l)}} \cot\left(\pi \frac{s_j^{(l)} + hr}{q}\right),$$

where $\sum_{j=1}^{q-1}{}^*$ means that the sum is extended over all values of j, for which

$$\left\{\frac{\theta j q}{r + a_j}\right\} \leq q^{-1}2^{m_1}$$

for either $\theta = 1$ or $\theta = -1$,

$$Q_1\left(\frac{r+a_l}{q}\right) = Q\left(\frac{r+a_l}{q}\right) - Q_0\left(\frac{r+a_l}{q}\right),$$

m_1 is a fixed positive integer.

(For the conclusion of the proof we let $m_1 \to \infty$).

The size of $Q\left(\frac{r+a_l}{q}\right)$ and also of $c_0\left(\frac{r+a_l}{q}\right)$ is essentially determined by Q_0, since Q_1 is small. The rest of the proof of the results of Theorem 2.4.6 proceeds in an analogous manner to the proof of Theorem 2.1.5.

2.5. Results Related to the Nyman–Beurling Criterion for the Riemann Hypothesis

According to the approach of Nyman–Beurling–Báez–Duarte (see [6, 16]) to the Riemann Hypothesis, the Riemann Hypothesis is true if and only if $\lim_{N\to\infty} d_N^2 = 0$, where

$$d_N^2 = \inf_{D_N} \frac{1}{2\pi} \int_{-\infty}^{\infty} \left|1 - \zeta\left(\frac{1}{2} + it\right) D_N\left(\frac{1}{2} + it\right)\right|^2 \frac{dt}{\frac{1}{4} + t^2} \qquad (2.5.1)$$

and the infimum is over all Dirichlet polynomials

$$D_N(s) := \sum_{n=1}^{N} \frac{a_n}{n^s}, \quad a_n \in \mathbb{C},$$

of length N (see [17]). It follows from [17] that under certain assumptions among all Dirichlet polynomials $D_N(s)$ the infimum in (2.5.1) is attained for $D_N(s) = V_N(s)$, where

$$V_N(s) := \sum_{n=1}^{N} \left(1 - \frac{\log n}{\log N}\right) \frac{\mu(n)}{n^s}. \qquad (2.5.2)$$

It thus is of interest to obtain an unconditional estimate for the integral in (2.5.1).

If we expand the square in (2.5.1) we obtain

$$d_N^2 = \inf_{D_N} \left(\int_{-\infty}^{\infty} \left(1 - \zeta\left(\frac{1}{2} + it\right) D_N\left(\frac{1}{2} + it\right) \right. \right.$$
$$\left. - \zeta\left(\frac{1}{2} - it\right) \overline{D}_N\left(\frac{1}{2} + it\right)\right) \frac{dt}{\frac{1}{4} + t^2}$$
$$\left. + \int_{-\infty}^{\infty} \left|\zeta\left(\frac{1}{2} + it\right)\right|^2 \left|D_N\left(\frac{1}{2} + it\right)\right|^2 \frac{dt}{\frac{1}{4} + t^2} \right).$$

The last integral evaluates as

$$\sum_{1 \leq h,k \leq N} a_h \bar{a}_k h^{-1/2} k^{-1/2} \int_{-\infty}^{\infty} \left|\zeta\left(\frac{1}{2} + it\right)\right|^2 \left(\frac{h}{k}\right)^{it} \frac{dt}{\frac{1}{4} + t^2}.$$

We have

$$b_{h,k} := \frac{1}{2\pi\sqrt{hk}} \int_{-\infty}^{\infty} \left|\zeta\left(\frac{1}{2} + it\right)\right|^2 \left(\frac{h}{k}\right)^{it} \frac{dt}{\frac{1}{4} + t^2}$$
$$= \frac{\log 2\pi - \gamma}{2}\left(\frac{1}{h} + \frac{1}{k}\right) + \frac{k-h}{2hk}\log\frac{h}{k} - \frac{\pi}{2hk}\left(V\left(\frac{h}{k}\right) + V\left(\frac{k}{h}\right)\right),$$

$$(2.5.3)$$

where

$$V\left(\frac{h}{k}\right) := \sum_{m=1}^{k-1} \left\{\frac{mh}{k}\right\} \cot\left(\frac{\pi mh}{k}\right)$$

is Vasyunin's sum (see [125]). From (2.5.2) and (2.5.3), we obtain

$$\int_{-\infty}^{\infty} \left|\zeta\left(\frac{1}{2} + it\right)\right|^2 \left|D_N\left(\frac{1}{2} + it\right)\right|^2 \frac{dt}{\frac{1}{4} + t^2}$$
$$= \sum_{1 \leq h,k \leq N} \mu(h)\mu(k)\left(1 - \frac{\log h}{\log N}\right)\left(1 - \frac{\log k}{\log N}\right)$$
$$\times \left[\frac{\log 2\pi - \gamma}{2}\left(\frac{1}{h} + \frac{1}{k}\right) + \frac{k-h}{2h\pi}\log\frac{h}{k} - \frac{\pi}{2hk}\left(V\left(\frac{h}{k}\right) + V\left(\frac{k}{h}\right)\right)\right],$$

where $h, k \in \mathbb{N}$, $k \geq 2$, $1 \leq h \leq k$, $(h, k) = 1$ (see [16, 62]).

It has been shown that Vasyunin's sum V is related to the cotangent sums

$$c_0\left(\frac{h}{k}\right) := \sum_{l=1}^{k-1} \frac{l}{k} \cot\left(\frac{\pi h l}{k}\right),$$

by

$$V\left(\frac{h}{k}\right) = -c_0\left(\frac{\bar{h}}{k}\right),\qquad\qquad(2.5.4)$$

where $h\bar{h} \equiv 1 \;(\mathrm{mod}\; k)$ and to the value at $s = 0$ or $s = 1$ of the Estermann zeta function by the functional equation of the *imaginary part* [45]. Namely

$$c_0\left(\frac{a}{q}\right) = \frac{1}{2}D_{sin}\left(0,\frac{a}{q}\right) = 2q\pi^{-2}D_{sin}\left(1,\frac{\bar{a}}{q}\right),$$

where for $x \in \mathbb{R}$, $\mathrm{Re}(s) > 1$, we have

$$D_{\sin}(s,x) := \sum_{n=1}^{\infty} \frac{d(n)\sin(2\pi n x)}{n^s},$$

and $a\bar{a} \equiv 1 \;(\mathrm{mod}\; q)$. If $x \in \mathbb{R} \setminus \mathbb{Q}$, then Wilton [128] showed that the convergence of the above series at $s = 1$ is equivalent to the convergence of

$$\sum_{n\geq 1}(-1)^n \frac{\log v_{n+1}}{v_n},$$

where u_n/v_n denotes the nth partial quotient of x. A basic ingredient in the papers [61, 64] have been the representations of Balazard, Martin in their papers [11, 12], of the function

$$g(x) := \sum_{l\geq 1} \frac{1 - 2\{lx\}}{l},$$

involving the Gauss transform from the theory of continued fractions as well as from the paper [72], by Marmi, Moussa and Yoccoz. One can show (see [23]) that $g(x)$ can also be written in the form

$$-\sum_{n\geq 1} \frac{d(n)}{\pi n}\sin(2\pi n x) = -\frac{1}{\pi}D_{sin}(1,x).$$

Maier and Rassias investigated a partial sum, in which k is kept fixed, namely (up to a constant depending only on n):

$$\sum_{n \in I} \mu(n) \left(1 - \frac{\log n}{\log N}\right) \frac{1}{n} V \left(\frac{n}{k}\right),$$

which coincides with

$$\sum_{n \in I} \mu(n) \left(1 - \frac{\log n}{\log N}\right) g \left(\frac{n}{k}\right)$$

(I is a suitable interval).

Maier and Rassias proved the following theorem.

Theorem 2.5.1. *Let*

$$0 < \delta < D/2, \quad k^{2\delta} \leq B \leq k^D,$$

where

$$k^{-\delta} \leq \eta \leq 1.$$

Then, there is a positive constant β depending only on δ and D such that

$$\sum_{Bk \leq n < (1+\eta)Bk} \mu(n) g \left(\frac{n}{k}\right) = O \left((\eta Bk)^{1-\beta}\right).$$

Corollary 2.5.2. *Let $\delta > 0$, $\epsilon > 0$ fixed, $k \geq N^\delta$. Then there is $\kappa > 0$ depending only on δ, ϵ such that*

$$\sum_{1 \leq n \leq N} \frac{\mu(n)}{n} \left(1 - \frac{\log n}{\log N}\right) V \left(\frac{n}{k}\right)$$

$$= \sum_{1 \leq n \leq k'} \frac{\mu(n)}{n} \left(1 - \frac{\log n}{\log N}\right) V \left(\frac{n}{k}\right) + O((N + k)^{1-\kappa}),$$

where $k' = k^{1+\epsilon}$.

Proof.   This follows from Theorem 2.5.4 by integration by parts.   □

They express the Möbius function by Vaughan's identity.

Lemma 2.5.3. *Let $w > 1$. For $n \in \mathbb{N}$ we have*

$$\mu(n) = c_1(n) + c_2(n) + c_3(n),$$

where

$$c_1(n) := \sum_{\substack{\alpha\beta\gamma=n \\ \alpha \geq w,\, \beta \geq w}} \mu(\gamma)c_4(\alpha)c_4(\beta),$$

for

$$c_4(\alpha) := - \sum_{\substack{(d_1,d_2)\,:\,d_1d_2=\alpha \\ d_1 \leq w}} \mu(d_1).$$

$$c_2(n) := \begin{cases} 2\mu(n) & \text{if } n \leq w, \\ 0 & \text{if } n > w, \end{cases}$$

and

$$c_3(n) := - \sum_{\substack{\alpha\beta\gamma=n \\ \alpha \leq w,\, \beta \leq w}} \mu(\alpha)\mu(\beta).$$

Proof. We introduce the generating Dirichlet series

$$\sum_{n=1}^{+\infty} c_1(n)n^{-s} = \frac{1}{\zeta(s)}E(s)^2,$$

$$\sum_{n=1}^{+\infty} c_2(n)n^{-s} = 2M_w(s),$$

$$\sum_{n=1}^{+\infty} c_3(n)n^{-s} = -M_w(s)^2\zeta(s),$$

where

$$M_w(s) := \sum_{n \leq w} \mu(n)n^{-s} \quad \text{and} \quad E(s) := \zeta(s)M_w(s) - 1.$$

The proof follows by comparison of the coefficients of these Dirichlet series.   □

They now obtain

$$\sum_{Bk \leq n < (1+\eta)Bk} \mu(n)g\left(\frac{n}{k}\right) = \Sigma_1 + \Sigma_2 + \Sigma_3,$$

where

$$\Sigma_i := \sum_{Bk \leq n < (1+\eta)Bk} c_i(n)g\left(\frac{n}{k}\right).$$

From the definitions for c_1 and c_4 in Lemma 2.5.3, we obtain:

$$\Sigma_1 = \sum_{\substack{s \geq w,\, t \geq w \\ Bk \leq st\gamma < (1+\eta)Bk}} \mu(\gamma) \sum(s,t),$$

where

$$\sum(s,t) := \sum_{\substack{d_1 d_2 = s \\ d_1 \leq w}} \mu(d_1) \sum_{\substack{e_1 e_2 = t \\ e_1 \leq w}} \mu(e_1)g\left(\frac{d_1 d_2 e_1 e_2 \gamma}{k}\right).$$

We now choose $w = (Bk)^{2\delta_0}$, for $\delta_0 > 0$ fixed, depending only on δ, η and D. The subsequent steps are all valid, if δ_0 is sufficiently small. They partition the sum Σ_1 as follows:

$$\Sigma_1 = \Sigma_{1,1} + \Sigma_{1,2}$$

with

$$\Sigma_{1,1} := \sum_{\substack{s \geq w,\, t \geq w \\ Bk \leq st\gamma < (1+\eta)Bk \\ st \geq (Bk)^{10\delta_0}}} \mu(\gamma) \sum(s,t)$$

and

$$\Sigma_{1,2} := \sum_{\substack{s \geq w,\, t \geq w \\ Bk \leq st\gamma < (1+\eta)Bk \\ st < (Bk)^{10\delta_0}}} \mu(\gamma) \sum(s,t).$$

The main idea for the treatment of these sums is the use of the antisymmetry of the function g. For details, see [67]. In the paper [68], Maier and Rassias consider sums similar to those in the paper [67]. Here they obtained explicit values for the exponent. They proved the following theorem.

Theorem 2.5.4. *Let $D \geq 2$. Let C be the number which is uniquely determined by*

$$C \geq \frac{\sqrt{5}+1}{2}, \quad 2C - \log C - 1 - 2\log 2 = \frac{1}{2}\log 2.$$

Let v_0 be determined by

$$v_0 \left(1 - \left(1 + 2 \log 2 \left(C + \frac{\log 2}{2} \right)^{-1} \right)^{-1} + 2 + \frac{4}{\log 2} C \right) = 2.$$

Let

$$z_0 := 2 - \left(2 + \frac{4}{\log 2} C \right) v_0.$$

Then, for all $\epsilon > 0$, we have

$$\sum_{k^D \leq n < 2k^D} \mu(n) g\left(\frac{n}{k} \right) \ll_\epsilon k^{D - z_0 + \epsilon}.$$

Burnol [26], improving on work of Báez-Duarte, Balazard, Landreau and Saias [6, 7] showed that

$$\liminf_{N \to \infty} d_N^2 \log N \geq \sum_{Re(\rho) = \frac{1}{2}} \frac{m(\rho)^2}{|\rho|^2},$$

where $m(\rho)$ denotes the multiplicity of the zero ρ. This lower bound is believed to be optimal and one expects that

$$d_N^2 \sim \frac{1}{\log N} \sum_{Re(\rho) = \frac{1}{2}} \frac{m(\rho)^2}{|\rho|^2}. \tag{*}$$

Under the Riemann Hypothesis, one has

$$\sum_{Re(\rho) = \frac{1}{2}} \frac{m(\rho)}{|\rho|^2} = 2 + \gamma - \log 4\pi, \tag{2.5.5}$$

where γ is the Euler–Mascheroni constant. Bettin, Conrey and Farmer [17] proved (*) under an additional assumption and also identified the Dirichlet polynomials A_N, for which the expected infimum in p. 70 is assumed. They proved (Theorem 1 of [17]) the following:

Let

$$V_N(s) := \sum_{n=1}^{N} \left(1 - \frac{\log n}{\log N} \right) \frac{\mu(n)}{n^s}.$$

If the Riemann hypothesis is true and if

$$\sum_{|Im(\rho)| \leq T} \frac{1}{|\zeta'(\rho)|^2} \ll T^{\frac{3}{2} - \delta}$$

for some $\delta > 0$, then

$$\frac{1}{2\pi} \int_{-\infty}^{\infty} \left| 1 - \zeta V_N \left(\frac{1}{2} + it \right) \right|^2 \frac{dt}{\frac{1}{4} + t^2} \sim \frac{2 + \gamma - \log 4\pi}{\log N}. \tag{2.5.6}$$

In the paper [66] Maier and Rassias investigated d_N^2 under an assumption contrary to the Riemann Hypothesis: There are exactly four non-trivial zeros off the critical line. They observed that non-trivial zeros off the critical line always appear as quadruplets. Indeed, if $\zeta(\rho) = 0$ for $\rho = \sigma + i\gamma$ with $1 > \sigma > \frac{1}{2}$, $\gamma > 0$, then from the functional equation

$$\Lambda(s) = \Lambda(1 - s), \tag{2.5.7}$$

where

$$\Lambda(s) := \pi^{-s/2} \Gamma\left(\frac{s}{2}\right) \zeta(s),$$

and the trivial relation $\zeta(\bar{s}) = \overline{\zeta(s)}$, we obtain that

$$\zeta(\sigma + i\gamma) = \zeta(1 - \sigma + i\gamma) = \zeta(\sigma - i\gamma) = \zeta(1 - \sigma - i\gamma) = 0.$$

They prove the following theorem.

Theorem 2.5.5. *Let $\sigma_0 > 1/2$, $\gamma_0 > 0$,*

$$\zeta(\sigma_0 \pm i\gamma_0) = \zeta(1 - \sigma_0 \pm i\gamma_0) = 0$$

and $\zeta(\sigma + i\gamma) \neq 0$ for all other $\sigma + i\gamma$ with $\sigma > 1/2$. Assume that

$$\sum_{|Im(\rho)| \leq T} \frac{1}{|\zeta'(\rho)|^2} \ll T^{\frac{3}{2} - \delta} \quad (T \to \infty),$$

for some $\delta > 0$. Then, there are real constants $A = A(\sigma_0, \gamma_0)$, $B = B(\sigma_0, \gamma_0)$ and $C = C(\sigma_0, \gamma_0)$ such that for all $\epsilon > 0$:

$$\frac{1}{2\pi} \int_{-\infty}^{\infty} \left| 1 - \zeta V_N \left(\frac{1}{2} + it \right) \right|^2 \frac{dt}{\frac{1}{4} + t^2}$$

$$= \frac{1}{(\log N)^2} \Big(A N^{2\sigma_0 - 1} \cos(2\gamma_0 \log N)$$

$$+ B N^{2\sigma_0 - 1} \sin(2\gamma_0 \log N) + C N^{2\sigma_0 - 1} \Big) \Big(1 + O(N^{\frac{1}{2} - \sigma_0 + \epsilon}) \Big).$$

2.6. The Maximum of Cotangent Sums for Rational Numbers in Short Intervals

In the paper [65], Maier and Rassias investigated the maximum of $\left|c_0\left(\frac{r}{b}\right)\right|$ for the values r/b in a short interval. They started with the following definition.

Definition 2.6.1. Let $0 < A_0 < 1$, $0 < C < 1/2$. For $b \in \mathbb{N}$, we set $\Delta := \Delta(b, c) = b^{-C}$ and

$$M(b, C, A_0) := \max_{A_0 b \leq r < (A_0 + \Delta)b} \left|c_0\left(\frac{r}{b}\right)\right|.$$

They proved the following results.

Theorem 2.6.2. *With Definition 2.6.1 let D satisfy $0 < D < \frac{1}{2} - C$. Then, we have for sufficiently large b:*

$$M(b, C, A_0) \geq \frac{D}{\pi} b \log b.$$

Theorem 2.6.3. *Let C be as in Theorem 2.5.5 and let D satisfy*

$$D > 2 - C - E, \text{ where } E \geq 0 \text{ is a fixed constant.}$$

Let B be sufficiently large. Then, it holds:

$$M(b, C, A_0) \leq \frac{D}{\pi} b \log b,$$

for all b with $B \leq b < 2B$ with at most B^E exceptions.

Basic for the proof of the above two theorems are the following preliminary lemmas, which are due to Bettin [15]. We recall the following definition.

Definition 2.6.4. For $x \in \mathbb{R}$, $\mathrm{Re}(s) > 1$, set

$$D_{\sin}(s, x) := \sum_{n \geq 1} \frac{d(n) \sin(2\pi n x)}{n^s}. \tag{*}$$

Lemma 2.6.5. *Let* $\langle a_0; a_1, a_2, \ldots \rangle$ *be the continued fraction expansion of* $x \in \mathbb{R}$. *Moreover, let* u_r/v_r *be the rth partial quotient of* x. *Then*

$$D_{sin}(1, x) = -\frac{\pi^2}{2} \sum_{l \geq 1} \frac{(-1)^l}{v_l} \left(\left(\frac{1}{\pi v_l} \right) + \psi \left(\frac{v_{l-1}}{v_l} \right) \right), \qquad (**)$$

whenever either of the two series (*), (**) *is convergent.*
If $x = \langle a_0; a_1, a_2, \ldots, a_r \rangle$ *is a rational number, then the range of summation of the series on the right is to be interpreted as* $1 \leq l \leq r$. *Here* ψ *is an analytic function satisfying*

$$\psi(x) = -\frac{\log(2\pi x) - \gamma}{\pi x} + O(\log x), \qquad (x \to 0).$$

Lemma 2.6.6.

$$c_0 \left(\frac{r}{b} \right) = \frac{1}{2} D_{sin} \left(0, \frac{r}{b} \right) = 2b \, \pi^{-2} D_{sin} \left(1, \frac{\bar{r}}{b} \right), \quad \text{where } r\bar{r} \equiv 1 (\text{mod } b).$$

Of crucial importance is the counting function defined below.

Definition 2.6.7. Let Δ be as in Definition 2.6.1 and $\Omega > 0$. We set

$$N(b, \Delta, \Omega) := \#\{r \; : \; A_0 \, b \leq r < (A_0 + \Delta)b, \; |\bar{r}| \leq \Omega \, b\}.$$

The localization of the multiplicative inverse $\bar{r}$ is accomplished by Fourier Analysis and estimates of Kloostermann sums. For the proof of Theorem 2.6.3, it is important to ensure that the influence of the terms $\psi \left(\frac{v_{l-1}}{v_l} \right)$ in Lemma 2.6.5 is small. This is achieved by the following Lemma.

Lemma 2.6.8. *Let*

$$\epsilon > 0, \; B \geq B(\epsilon), \; B < b \leq 2B.$$

For $1 \leq r < b$, $(r, b) = 1$, *let*

$$\frac{r}{b} = \langle 0; w_1, \ldots, w_s \rangle$$

be the continued fraction expansion of r/b *with partial fractions* u_i/v_i. *Then there are at most 3 values of* l *for which*

$$\frac{1}{v_l} \psi \left(\frac{v_{l-1}}{v_l} \right) \geq \log \log b$$

and at most one value of l, *for which*

$$\frac{1}{v_l} \psi \left(\frac{v_{l-1}}{v_l} \right) \geq \epsilon \log b.$$

Proof.   Let l_i $(i = 1, 2, 3, 4)$ be such that

$$\frac{1}{v_{l_i}} \psi \left(\frac{v_{l_i - 1}}{v_{l_i}} \right) \geq \log \log b.$$

Then

$$v_{l_1} \geq \log \log b, \ v_{l_2} \geq \exp(v_{l_1}) \geq \log b, \ v_{l_3} \geq \exp(v_{l_2}) \geq b,$$

$$v_{l_4} \geq \exp(v_{l_3}) \geq \exp(b),$$

in contradiction to $v_s \leq b$. In the same manner, we obtain from $v_{l_j} \geq \epsilon \log b$, $j = 1, 2$:

$$v_s \geq \exp(\exp((\log b)^\epsilon)) > b. \qquad \square$$

2.7. Open Problems

(1) Investigate the distribution of the cotangent sums $c_0(r/b)$ if r is fixed and b is variable or if both r and b are variable.

(2) Is it possible to obtain a sharper form of the asymptotics for the kth moments containing a second main term or an asymptotic expansion?

(3) Can one get sharp asymptotics for the kth moments of cotangent sums that are uniform in k?

(4) Can one get further results for other sequences of rational numbers, like rational numbers with smooth numerators or denominators or square numbers as numerators and denominators?

(5) Can one obtain results on the integral expression

$$\frac{1}{2\pi} \int_{-\infty}^{\infty} \left| 1 - \zeta V_N \left(\frac{1}{2} + it \right) \right|^2 \frac{dt}{\frac{1}{4} + t^2}$$

assuming the existence of more than four zeros of the zeta function off the critical line or an infinite configuration of such zeros?

(6) What results can be obtained if the maximum of $c_0(r/b)$ is taken over the values of r as well as over the values of b in short intervals for the fraction r/b?

Bibliography

[1] H. L. Alder, A generalization of the Euler ϕ–function, *Amer. Math. Monthly* **65** (1958) 690–692.

[2] E. Alkan, Distribution of averages of Ramanujan sums, *Ramanujan J.* **29** (2012) 385–408.

[3] E. Alkan, Ramanujan sums and the Burgess zeta function, *Int. J. Number Theory* **8** (2012) 2069–2092.

[4] T. M. Apostol, *Introduction to Analytic Number Theory*, Springer, 1976.

[5] J. S. Auli, A. Bayad, M. Beck, Reciprocity theorems for Bettin–Conrey sums. *Acta Arith.* **181**(4) (2017) 297–319.

[6] L. Báez-Duarte, M. Balazard, B. Landreau, E. Saias, Notes sur la fonction ζ de Riemann. III [Notes on the Riemann ζ-function. III], *Adv. Math.* **149**(1) (2000) 130–144 (in French).

[7] L. Báez-Duarte, M. Balazard, B. Landreau, E. Saias, Étude de l'autocorrelation multiplicative de la fonction 'partie fractionnaire', [Study of the multiplicative autocorrelation of the fractional part function], *Ramanujan J.* **9**(1–2) (2005) 215–240; arxiv math.NT/0306251 (in French).

[8] B. Bagchi, On Nyman, Beurling and Baez-Duarte's Hilbert space reformulation of the Riemann hypothesis, *Proc. Indian Acad. Sci. Math.* **116**(2) (2006) 137–146.

[9] R. C. Baker, The square-free divisor problem II, *Quart. J. Math. (Oxford)* (2) **47** (1996) 133–146.

[10] R. Balasubramanian, J. B. Conrey, D. R. Heath-Brown, Asymptotic mean square of the product of the Riemann zeta-function and a Dirichlet polynomial, *J. Reine Angew. Math.* **357** (1985) 161–181.

[11] M. Balazard, B. Martin, Comportement local moyen de la fonction de Brjuno [Average local behavior of the Brjuno function], *Fund. Math.* **218**(3) (2012) 193–224 (in French).

[12] M. Balazard, B. Martin, Sur l'autocorrélation multiplicative de la fonction "partie fractionnaire" et une fonction définie par J. R. Wilton, preprint (2013), arXiv:1305.4395v1.

[13] A. Bayad, M. Goubi, Reciprocity formulae for generalized Dedekind-Vasyunin-cotangent sums, *Math. Methods Appl. Sci.* **42**(4) (2019) 1082–1098.

[14] S. Bettin, A generalization of Rademacher's reciprocity law, *Acta Arithmetica* **159**(4) (2013) 363–374.

[15] S. Bettin, On the distribution of a cotangent sum, *Int. Math. Res. Notices* (2015); doi: 10.1093/imrn/rnv036.

[16] S. Bettin, B. Conrey, Period functions and cotangent sums, *Algebra Number Theory* **7**(1) (2013) 215–242.

[17] S. Bettin, J. B. Conrey, D. W. Farmer, An optimal choice of Dirichlet polynomials for the Nyman-Beurling criterion, *Proc. Steklov Inst. Math.* **280**(2 Suppl) (2013) 30–36 (in honor of A.A. Karatsuba).

[18] S. Bettin, High moments of the Estermann function, *Algebra Number Theory*, **13**(2) (2019) 251–300.

[19] S. Bettin, S. Drappeau, Partial sums of the cotangent function, *J. Théorie Nombres Bordeaux*, 32(1) (2020) 217–230.

[20] K. Bibak, B. M. Kapron, V. Srinivasan, R. Tauraso, L. Tóth, Restricted linear congruences, *J. Number Theory* **171** (2017) 128–144.

[21] J. Bourgain, N. Watt, Mean square of zeta function, circle problem and divisor problem revisited, preprint (2017); https://arxiv.org/abs/1709.04340.

[22] A. Brauer, Lösung der Aufgabe 30, *Jber. Deutsch. Math.-Verein* **35** (1926) 92–94.

[23] R. de la Bretèche, G. Tenenbaum, Séries trigonométriques à coefficients arithmétiques, *J. Anal. Math.* *92* (2004) 1–79.

[24] R. W. Bruggeman, Eisenstein series and the distribution of Dedekind sums, *Math. Z.* **202** (1989) 181–198.

[25] R. W. Bruggeman, Dedekind sums and Fourier coefficients of modular forms, *J. Number Theory* **36** (1990) 289–321.

[26] J. F. Burnol, A lower bound in an approximation problem involving the zeros of the Riemann zeta function, *Advances in Math.* **170** (2002) 56–70.

[27] E. Cohen, Rings of arithmetic functions. II: The number of solutions of quadratic congruences, *Duke Math. J.* **21** (1954) 9–28.

[28] E. Cohen, A class of arithmetical functions, *Proc. Nat. Acad. Sci. USA* **41** (1955) 939–944.

[29] E. Cohen, Representations of even functions (mod r), I. Arithmetical identities, *Duke Math. J.* **25** (1958) 401–421.

[30] E. Cohen, Arithmetical functions associated with the unitary divisors of an integer, *Math. Z.* **74** (1960) 66–80.

[31] H. Cohen, *Number Theory, Vol. II. Analytic and Modern Tools*, Graduate Texts in Mathematics, Vol. 240, Springer, 2007.

[32] N. Derevyanko, K. Kovalenko, M. Zhukovskii, On a category of cotangent sums related to the Nyman–Beurling criterion for the Riemann hypothesis, In *Trigonometric Sums and Their Applications*, Springer, 2020, pp. 1–28.

[33] P. D. T. A. Elliott, *Probabilistic Number Theory I: Mean-Value Theorems*, Springer, New York, 1979.

[34] F. Erwe, Differential-und Integralrechnung II, *Bibl. Inst. Mannheim*, 1962.

[35] T. Estermann, On the representation of a number as the sum of two products, *Proc. London Math. Soc.* **31**(2) (1930) 123–133.

[36] S. Finch, G. Martin, P. Sebah, Roots of unity and nullity modulo n, *Proc. Amer. Math. Soc.* **138** (2010) 2729–2743.

[37] E. Fouvry, Ph. Michel, Sur certaines sommes d'exponentielles sur les nombres premiers, *Ann. Sci. École Norm. Sup. (4)*, **31** (1998) 93–130.

[38] G. Harman, *Metric Number Theory*, Oxford University Press, Oxford, 1998.

[39] P. Haukkanen, L. Tóth, An analogue of Ramanujan's sum with respect to regular integers (mod r), *Ramanujan J.* **27** (2012) 71–88.

[40] D. Hensley, *Continued Fractions*, World Scientific, Singapore, 2006.

[41] J. Herzog, P. R. Smith, Lower bounds for a certain class of error functions, *Acta Arith.* **60** (1992) 289–305.

[42] L. K. Hua, *Introduction to Number Theory*, Springer, 1982.

[43] M. N. Huxley, Exponential sums and lattice points III., *Proc. London Math. Soc.* **87** (2003) 591–609.

[44] S. Ikeda, I. Kiuchi, K. Matsuoka, Sums of products of generalized Ramanujan sums, *J. Integer Seq.* **19** (2016) Article 16.2.7, 22 pp.

[45] M. Ishibashi, The value of the Estermann zeta function at $s = 0$, *Acta Arith.* **73**(4) (1995) 357–361.

[46] M. Ishibashi, $\mathbb{Q}$-linear relations of special values of the Estermann zeta function, *Acta Arith.* **86**(3)(1998) 239–244.

[47] H. Iwaniec, On the mean values for Dirichlet's polynomials and the Riemann zeta function, *J. London Math. Soc.* **22**(2)(1980) 39–45.

[48] H. Iwaniec, E. Kowalski, Analytic Number Theory, *American Mathematical Society Colloqium*, Vol. 53, Providence, RI, 2004.

[49] K. R. Johnson, Unitary analogs of generalized Ramanujan sums, *Pacific J. Math.* **103** (1982) 429–432.

[50] G. Jones, P. Kester, L. Martirosyan, P. Moree, L. Tóth, B. White, B. Zhang, Coefficients of (inverse) unitary cyclotomic polynomials, *Kodai Math. J.* **43** (2020) 325–338.

[51] V. S. Joshi, Order-free integers (mod m), In *Number Theory* (Mysore, 1981), Lecture Notes in Mathematics, Vol. 938, Springer, 1982, pp. 93–100.

[52] I. Kiuchi, On an exponential sum involving the arithmetic function $\sigma_\alpha(n)$, *Math. J. Okayama Univ.* **29** (1987) 193–205.

[53] I. Kiuchi, Sums of averages of gcd-sum functions, *J. Number Theory* **176** (2017) 449–472.

[54] I. Kiuchi, Sums of averages of generalized Ramanujan sums, *J. Number Theory* **180** (2017) 310–348.

[55] I. Kiuchi, On sums of averages of generalized Ramanujan sums, *Tokyo J. Math.* **40** (2017) 255–275.

[56] A. Klenke, *Probability Theory*, Springer, Berlin, 2006.

[57] Y. Li, D. Kim, Menon-type identities with additive characters, *J. Number Theory* **192** (2018) 373–385.

[58] V. A. Liskovets, A multivariate arithmetic function of combinatorial and topological significance, *Integers* **10** (2010) 155–177.

[59] L. G. Lucht, A survey of Ramanujan expansions, *Int. J. Number Theory* **6** (2010) 1785–1799.

[60] H. Maier, M. Th. Rassias, The rate of growth of moments of certain cotangent sums, *Aequationes Math.*, **90**(3) (2016) 581–595.

[61] H. Maier, M. Th. Rassias, The order of magnitude for moments for certain cotangent sums, *J. Math. Anal. Appl.* **429**(1) (2015) 576–590.

[62] H. Maier, M. Th. Rassias, Generalizations of a cotangent sum associated to the Estermann zeta function, *Commun. Contemp. Math.* **18**(1) (2016) 89 pp. doi: 10.1142/S0219199715500789.

[63] H. Maier, M. Th. Rassias, Asymptotics for moments of certain cotangent sums, *Houston J. Math.* **43**(1) (2017) 207–222.

[64] H. Maier, M. Th. Rassias, Asymptotics for moments of certain cotangent sums for arbitrary exponents, *Houston J. Math.* **43**(4) (2017) 1235–1249.

[65] H. Maier, M. Th. Rassias, The maximum of cotangent sums related to Estermann's zeta function in rational numbers in short intervals, *Appl. Anal. Discrete Math.* **11** (2017) 166–176.

[66] H. Maier, M. Th. Rassias, On the size of an expression in the Nyman–Beurling–Baez–Duarte criterion for the Riemann Hypothesis, *Canadian Mathematical Bulletin*, **61**(3) (2018) 622-627.

[67] H. Maier, M. Th. Rassias, Estimates of sums related to the Nyman-Beurling criterion for the Riemann Hypothesis, *J. Number Theory*, **188**(2018) 96–120.

[68] H. Maier, M. Th. Rassias, Explicit estimates of sums related to the Nyman-Beurling criterion for the Riemann Hypothesis, *J. Funct. Anal.* **276**(2019) 3832–3857.

[69] H. Maier, M. Th. Rassias, Distribution of a cotangent sum related to the Nyman-Beurling criterion for the Riemann Hypothesis, *Appl. Math. Comput.* **363**(15)(2019); https://doi.org/10.1016/j.amc.2019.124589.

[70] H. Maier, M. Th. Rassias, Cotangent sums related to the Riemann Hypothesis for various shifts of the argument, *Canadian Math. Bull.* **63**(3) (2020) 522-535.

[71] H. Maier, M. Th. Rassias, A. Raigorodskii, The maximum of cotangent sums related to the Nyman-Beurling criterion for the Riemann Hypothesis, In *Trigonometric Sums and their Applications*, Springer, 2020, pp. 149–158.

[72] S. Marmi, P. Moussa, J.-C. Yoccoz, The Brjuno functions and their regularity properties, *Commun. Math. Phys.* **186** (1997) 265–293.

[73] P. J. McCarthy, Regular arithmetical convolutions, *Portugal. Math.* **27** (1968) 1–13.

[74] P. J. McCarthy, The number of restricted solutions of some systems of linear congruences, *Rend. Sem. Mat. Univ. Padova* **54** (1975) 59–68.

[75] P. J. McCarthy, *Introduction to Arithmetical Functions*, Springer, 1986.

[76] A. D. Mednykh, R. Nedela, Enumeration of unrooted maps with given genus, *J. Combin. Theory, Ser. B* **96** (2006) 706–729.

[77] P. K. Menon, On the sum $\sum (a - 1, n)[(a, n) = 1]$, *J. Indian Math. Soc. (N.S.)* **29** (1965) 155–163.

[78] H. L. Montgomery, *Early Fourier Analysis*, Pure and Applied Undergraduate Texts, Vol. 22, American Mathematical Society, Providence, RI, 2014.

[79] H. L. Montgomery, R. C. Vaughan, *Multiplicative Number Theory, I. Classical Theory*, Cambridge Studies in Advanced Mathematics, Vol. 97, Cambridge University Press, 2007.

[80] P. Moree, L. Tóth, Unitary cyclotomic polynomials, *Integers* **20** (2020), Paper No. A65, 21 pp.

[81] K. Motose, Ramanujan's sums and cyclotomic polynomials, *Math. J. Okayama Univ.* **47** (2005) 65–74.

[82] K. V. Namboothiri, Certain weighted averages of generalized Ramanujan sums, *Ramanujan J.* **44** (2017) 531–547.

[83] W. Narkiewicz, On a class of arithmetical convolutions, *Colloq. Math.* **10** (1963) 81–94.

[84] W. Narkiewicz, *Number Theory*, World Scientific, Singapore, 1983.

[85] M. B. Nathanson, *Additive Number Theory. The Classical Bases*, Graduate Texts in Mathematics, Vol. 164, Springer, 1996.

[86] C. A. Nicol, Some formulas involving Ramanujan sums, *Canad. J. Math.* **14** (1962) 284–286.

[87] C. A. Nicol, H. S. Vandiver, A von Sterneck arithmetical function and restricted partitions with respect to a modulus, *Proc. Natl. Acad. Sci. USA* **40** (1954) 825–835.

[88] K. Prachar, *Primzahlverteilung*, Springer, Berlin, 1957.

[89] A. V. Prokhorov, Borel–Cantelli lemma, In *Encyclopedia of Mathematics* M. Hazewinkel (ed.), Springer, New York, 2001.

[90] H. Rademacher, Aufgabe 30, *Jber. Deutsch. Math.–Verein* **34** (1925) 158.

[91] A. Raigorodskii, M. Th. Rassias (eds.), *Trigonometric Sums and their Applications*, Springer, 2020.

[92] S. Ramanujan, On certain trigonometric sums and their applications in the theory of numbers, *Trans. Cambridge Philos. Soc.* **22** (1918) 259–276; Collected Papers, Cambridge **21** (1927) 179–199.

[93] M. Ram Murty, Ramanujan series for arithmetic functions, *Hardy–Ramanujan J.* **36** (2013) 21–33.

[94] M. Th. Rassias, Analytic investigation of cotangent sums related to the Riemann zeta function, Doctoral Dissertation, ETH-Zürich, Switzerland, 2014.

[95] M. Th. Rassias, On a cotangent sum related to zeros of the Estermann zeta function, *Appl. Math. Comput.* **240**(2014) 161–167.

[96] D. Rearick, A linear congruence with side conditions, *Amer. Math. Monthly* **70** (1963) 837–840.

[97] N. Robles, A. Roy, Moments of averages of generalized Ramanujan sums, *Monatsh. Math.* **182** (2017) 433–461.

[98] A. M. Rockett, P. Szüsz, *Continued Fractions*, World Scientific, Singapore, 1992.

[99] W. Rudin, *Real and Complex Analysis*, McGraw-Hill, New York, 1966.

[100] W. Schwarz, J. Spilker, *Arithmetical Functions*, London Mathematical Society Lecture Note Series, Vol. 184, Cambridge University Press, 1994.

[101] R. Sivaramakrishnan, *Classical Theory of Arithmetic Functions*, Monographs and Textbooks in Pure and Applied Mathematics, Vol. 126, Marcel Dekker, 1989.

[102] R. Šleževičiene, J. Steuding, On the zeros of the Estermann zeta-function, *Integral Transforms Special Functions*, **13** (2002) 363–371.

[103] N. J. A. Sloane, *The On-Line Encyclopedia of Integer Sequences.* http://oeis.org.

[104] J. Spilker, Eine einheitliche Methode zur Behandlung einer linearen Kogruenz mit Nebenbedingungen, *Elem. Math.* **51** (1996) 107–116.

[105] M. V. Subbarao, A note on the arithmetic functions $C(n,r)$ and $C^*(n,r)$, *Nieuw Arch. Wisk.* **14**(3) (1966) 237–240.

[106] D. Suryanarayana, A property of the unitary analogue of Ramanujan's sum, *Elem. Math.* **25** (1970) 114.

[107] D. Suryanarayana, V. Siva Rama Prasad, The number of k-free divisors of an integer, *Acta Arith.* **17** (1971) 345–354.

[108] L. Tóth, Remarks on generalized Ramanujan sums and even functions, *Acta Math. Acad. Paedagog. Nyházi. (N.S.)* **20** (2004) 233–238.

[109] L. Tóth, Regular integers (mod n), *Annales Univ. Sci. Budapest., Sect. Comp.* **29** (2008) 263–275.

[110] L. Tóth, On the bi-unitary analogues of Euler's arithmetical function and the gcd-sum function, *J. Integer Seq.* **12** (2009), Article 09.5.2, 10 pp.

[111] L. Tóth, Some remarks on Ramanujan sums and cyclotomic polynomials, *Bull. Math. Soc. Sci. Math. Roumanie (N.S.)* **53**(101) (2010) 277–292.

[112] L. Tóth, Sums of products of Ramanujan sums, *Ann. Univ. Ferrara* **58** (2012) 183–197.

[113] L. Tóth, Some remarks on a paper of V. A. Liskovets, *Integers* **12** (2012) 97–111.

[114] L. Tóth, Counting solutions of quadratic congruences in several variables revisited, *J. Integer Seq.* **17** (2014), Article 14.11.6, 23 pp.

[115] L. Tóth, Averages of Ramanujan sums: Note on two papers by E. Alkan, *Ramanujan J.* **35** (2014) 149–156.

[116] L. Tóth, Multiplicative arithmetic functions of several variables: a survey, In *Mathematics Without Boundaries, Surveys in Pure Mathematics.* Th. M. Rassias, P. Pardalos (eds.), Springer, New York, 2014, pp. 483–514.

[117] L. Tóth, Menon-type identities concerning Dirichlet characters, *Int. J. Number Theory* **14** (2018) 1047–1054.

[118] L. Tóth, Ramanujan expansions of arithmetic functions of several variables, *Ramanujan J.* **47** (2018) 589–603.

[119] L. Tóth, Expansions of arithmetic functions of several variables with respect to certain modified unitary Ramanujan sums, *Bull. Math. Soc. Sci. Math. Roumanie* **61 (109)** (2018) 213–223.

[120] L. Tóth, Menon-type identities concerning additive characters, *Arab. J. Math.* **9** (2020) 697–705.

[121] L. Tóth, P. Haukkanen, The discrete Fourier transform of r-even functions, *Acta Univ. Sapientiae, Math.* **3** (2011) 5–25.

[122] N. Ushiroya, Ramanujan-Fourier series of certain arithmetic functions of two variables, *Hardy-Ramanujan J.* **39** (2016) 1–20.

[123] R. Vaidyanathaswamy, The theory of multiplicative arithmetic functions, *Trans. Amer. Math. Soc.* **33** (1931) 579–662.

[124] I. Vardi, Dedekind sums have a limiting distribution, *Int. Math. Res. Notices* 1(1993) 1–12.

[125] V. I. Vasyunin, On a biorthogonal system associated with the Riemann hypothesis, *Algebra i Analiz* **7**(3) (1995) 118–135 (in Russian); English translation in *St. Petersburg Math. J.* **7**(3) (1996) 405–419.

[126] A. Weil, On some exponential sums, *Proc. Natl. Acad. Sci. USA*, **34** (1948) 204–207.

[127] A. Weil, *Sur les courbes algébriques et les variétés qui s'en déduisent*, Paris, Hermann, 1948.

[128] J. R. Wilton, An approximate functional equation with applications to a problem of Diophantine approximation, *J. Reine Angew. Math. (Crelle's J.)*, **169** (1933) 219–237.

[129] D. Zagier, *Quantum modular forms*, In: *Quanta of Maths*, Clay Mathematics Proceedings 11, American Mathematical Society, Providence, RI, 2010, pp. 659–675.

[130] X.-P. Zhao, Z.-F. Cao, Another generalization of Menon's identity, *Int. J. Number Theory* **13** (2017) 2373–2379.

Index

Printed in the United States
by Baker & Taylor Publisher Services